A Guide To U.S. Aircraft Noise Regulatory Policy

Sanford Fidell · Vincent Mestre

A Guide To U.S. Aircraft Noise Regulatory Policy

Springer

Sanford Fidell
Woodland Hills, CA, USA

Vincent Mestre
Port Townsend, WA, USA

ISBN 978-3-030-39907-8 ISBN 978-3-030-39908-5 (eBook)
https://doi.org/10.1007/978-3-030-39908-5

This Springer imprint is published by the registered company Springer Nature Switzerland AG
The registered company address is: Gewerbestrasse 11, 6330 Cham, Switzerland

Preface

The World War II generation of acousticians, engineers, and social scientists who laid the technical foundations for aircraft noise regulation and management in the United States (among others, Leo Beranek, Richard Bolt, Per Bruel, Kenneth Eldred, Harvey Fletcher, William Galloway, Harvey Hubbard, Karl Kryter, Alan Marsh, Theodore Schultz, S. S. Stevens, Michael Smith, Louis Sutherland, Henning von Gierke, Robert Young, and Eberhard Zwicker) are now gone, and the ranks of the succeeding generation are thinning. Few of today's regulators, airport staff, consultants, and researchers can recall the details of yesterday's technical debates and policy analyses on which today's regulatory practices have been built. This monograph is intended to provide perspective for informed contemporary discussion of aviation noise regulation and management in the United States.

Community opposition to aircraft noise has been commonplace in airport-vicinity communities since the introduction of commercial jet service in the late 1950s. Aviation noise was not widely recognized as a societal problem in the United States, however, until a strong environmental ethos arose in the 1960s, and since requirements for disclosing and mitigating aircraft noise impacts of the National Environmental Policy Act and similar state legislation came into effect in the 1970s. Subsequent worldwide disputes between communities and airports over the effects of changes in aircraft operations and expansion of airport infrastructure have often been highly contentious, often to the point of straining relations among local governments, airports, airlines, and community residents; and occasionally to the extent of intense political controversy, lengthy litigation, and even mass protest.

Scientific understanding of aircraft noise and its effects, not to mention regulatory policy, international technical consensus standards, aviation technology, and public opinion have all evolved on separate timelines. U.S. policies supporting generous federal subsidies[1] for the air transportation industry since the 1920s are now being questioned, while the underpinnings of decades-old aircraft noise regulatory practice are being re-examined on technical and political grounds. For example, a 40-member Quiet Skies Caucus in the U.S. Congress incorporated numerous aircraft noise-related requirements in the FAA Reauthorization Act of 2018.[2] Due to a lack of full appreciation for the context in which aviation noise regulation has evolved in the United States, this re-examination is not always well-informed and technically reasonable. Further, near-term developments such as increased concentrations of flights in very narrow corridors many miles from airports, low altitude autonomous flight operations in urban settings, and overland supersonic flight risk spreading aircraft noise controversies well beyond the immediate environs of airports.

Productive discussion of aircraft noise regulatory policy is unlikely without an informed understanding of its origins and development. Examination of the data and assumptions on which noise regulations are based, and alternate interpretations of them, are rarely part of heated aircraft noise controversies, however. Instead, many such controversies revolve around perennial issues and familiar positions that have been repeatedly raised in differing contexts.

For example, the rationale for disclosure of predicted aircraft noise effects mandated by the National Environmental Policy Act ("NEPA") was developed in an era when the charter of the U.S. Federal Aviation Administration ("FAA") included promotion of civil aviation. Congress rescinded FAA's responsibility and authority for promoting civil aviation in 1996. National noise policies have not yet been revised to reflect FAA's new charter, and only recently have preliminary discussions of policy revisions begun. The agency has not yet revised its definition of "significant" noise impact, however, nor its views about land uses that it considers compatible with continued operation and expansion of airports. Policies based on obsolete goals and erroneous technical information have been retained simply because they have survived purely judicial (i.e., non-technical) review.

This monograph attempts to cast enough light on the origins and assumptions of aircraft noise regulation in the United States to assist in its continuing improvement. Although the focus of the monograph is aircraft noise regulation, understanding of underlying issues sometimes warrants discussion of related practical matters and technical literature. The monograph is not, however, intended as a tutorial on aircraft noise measurement and noise effects, and thus lacks detailed discussions of these topics. Readers interested in these and other basics may find them in Mestre et al. (2011) and Crocker (1997), among other places.

Woodland Hills, USA Sanford Fidell
Port Townsend, USA Vincent Mestre

Contents

List of Figures

Chapter 1
Introduction

A brief consideration of the usual definition of noise as "unwanted sound" is a convenient starting point. In the context of community exposure to aircraft noise, "unwanted sound" is a rather elliptical definition of noise, because it lacks an acknowledgment that aircraft noise is sound that someone *else* —that is, someone other than an unwilling listener—considers too inconvenient (costly, inefficient, or statutorily unnecessary) to control. The "someone *else*" qualification is useful because it draws attention to a basic distinction between human reactions to noise exposure and those of sound level meters. Sound level meters are designed to respond solely to miniscule air pressure changes at a microphone diaphragm. Individual and community reactions to sounds, however, are sensitive to many non-acoustic factors as well. Ignoring such additional factors by treating regulation strictly as a matter of acoustic analysis and engineering is a recipe for ineffective regulation.

The "someone *else*" qualification also helps to explain the inescapably political and economic aspects of airport noise controversies. The "us/them", adversarial nature of reactions to aircraft noise provokes special interest concerns, while elevating such controversies beyond the reach of local resolution. The "someone *else*" distinction also helps to clarify the institutional nature of aircraft noise controversies, as well as their extension beyond immediate, airport-vicinity residents to wider political and commercial realms, and to the national legislative, regulatory, and judicial establishments.

1.1 Varied Influences on Development of U.S. Aircraft Noise Regulation

Civil aircraft noise regulation in the United States has developed episodically over decades, in concert with the growth of commercial and private aviation, punctuated by occasional changes in legislative direction and judicial decisions. Aircraft noise regulation has also been influenced by advances in technical understandings

S. Fidell and V. Mestre, *A Guide To U.S. Aircraft Noise Regulatory Policy*,
https://doi.org/10.1007/978-3-030-39908-5_1

of community response to transportation noise; by changes in policy positions; by improvements in acoustic instrumentation; and by development of international standards and agreements. Although the fundamental goal of all regulation is to balance conflicting societal interests, a lasting, long-term balance between satisfying public demand for air transportation services on the one hand, and for habitable residential neighborhoods near airports on the other, has proved difficult to achieve.

In retrospect, past aviation noise regulatory policy has not often been successful in accomplishing its own goals,[3] much less in equitably balancing conflicting public, institutional, and private interests.[4] In many cases, intense controversies over increases in aircraft noise exposure have delayed or prevented construction of desired additional airport capacity. In other cases, lenient aircraft noise regulation has encouraged construction of expensive excess capacity at a few over-built airports.[5]

Both community and aviation industry interests have felt aggrieved by aviation noise regulation for decades. Communities commonly construe aircraft noise regulatory policy as intended to protect airport operations from limitations imposed to reduce community impacts, while airport managements tend to believe that airport noise policy favors consideration of community more than airport interests. As discussed further in Sect. 3.10.2, such beliefs reflect perspectives and expectations of both parties.

The regulatory balance sought since the 1950s has been complicated by dramatic growth in the size of the commercial jet aircraft fleet and changes in the character of aircraft engine noise. Air traffic increased enormously following industry deregulation in 1978, due to factors such as the introduction of more fuel-efficient and longer-range aircraft; the shift from city-pair routes to hub-and-spoke networks; the introduction (and retirement) of small regional jets; and the success of low- and ultra-low-cost alternatives to the business models of legacy carriers. Regulatory provisions have been influenced by industry lobbying for favorable treatment; by changes in the charter of the primary regulatory agency; and by public awareness of and distaste for environmental pollution.

A full understanding of policies affecting measurement, management, and regulation of aircraft noise requires an appreciation of many short- and long-term matters. These include:

- Federal legislation, international standards, litigation, and judicial decisions;
- Technical understandings gained from research on individual and community response to transportation noise;
- Jet engine development and design of civil air transports;
- Field measurement and software modeling of aircraft noise exposure;
- Conversion of airports from small municipal utilities exploiting America's patrimony of World War II military airfields into much higher capacity hubs; and
- Efforts to develop State and local land use policies for noise-compatible land uses in the vicinity of airports.

1.2 Tacit Assumptions of Current Aircraft Noise Regulatory Policy

Some of the tacit assumptions that underlie the rationale for aircraft noise regulation that are now undergoing re-examination include the following:

(1) Airports remain the best and highest use of expensive urban land;
(2) Societal needs for air transportation services outweigh needs for habitable neighborhoods;
(3) Adverse community response to aircraft noise is reliably predictable from long-term, cumulative noise exposure alone;
(4) A generic ("one-size-fits-all") dose–response relationship between cumulative exposure to *all* forms of transportation noise and the prevalence of a consequential degree of annoyance supports useful predictions of adverse public response to aircraft noise in all communities nationwide; and
(5) Current definitions of "significant" noise exposure and thresholds of eligibility for disclosure and mitigation of aircraft noise are objective ("fact- and science-based") and well-supported.

1.3 Perennial Issues

Despite major changes in the character of residential aircraft noise exposure and scientific understanding of its effects, questions and policy answers of aircraft noise regulation have changed little in six decades. Perennial questions include "What is the most apt indicator of aircraft noise effects on residential populations?"[6]; "What is the most accurate and technically defensible way to predict residential effects of aircraft noise exposure?"; "How much noise is too much noise?"; "How can you tell?"; "Who gets to decide?"; and "What can be done about aircraft noise?" The persistence of these issues is somewhat surprising, given the major changes in quantity and quality of aircraft noise exposure over the decades, and improved technical understandings of community response to aviation noise.

The "aircraft noise problem" is still widely regarded as one of measurement—devising a "magic bullet" scale to enable accurate and precise characterization of exposure—rather than one of systematic understanding of aircraft noise effects. Following Schultz's (1978) *tour de force* meta-analysis of social survey findings (see Sect. 3.2.2), the primary indication of adverse community response to aircraft noise switched in the 1980s from complaints (a behavior) to annoyance (an attitude). Numbers of aircraft operations at large civil airports that were unimaginable in the 1950s are now commonplace. Nonetheless, the only measures for mitigating excessive aircraft noise exposure acknowledged today are the same ones recognized decades ago. Dose–response analyses are still based (1) on naïve and simplistic notions of the sort favored in the 1970s, that people respond to aircraft noise in much the same was as do sound level meters, and (2) on generic, atheoretical, and purely correlational statistical methods.

Today's thresholds of regulatory "significance" of aircraft noise exposure are identical to those adopted in the 1950s. These thresholds were originally recommended to keep complaints about military aircraft noise from residents of airbase housing during the Cold War era to manageable numbers. When the goal of regulation shifted from managing complaints at military airbases to protecting civil residential populations from exposure to highly annoying aircraft noise, no meaningful effort was made to re-examine earlier recommendations about the significance of aircraft noise exposure. (Instead, the earlier units of measurement were simply mathematically converted into later units.) Likewise, no serious effort was made to re-examine policy positions on noise exposure regulatory thresholds after FAA lost its Congressional charter to foster the growth of civil aviation.

In the context of individual airport/community controversies, regulatory answers to the perennial questions have been for decades, and remain today, as follows:

- The percentage of the residential population that describes itself as highly annoyed by long-term, cumulative aircraft noise exposure is the most apt measure of the effects of aircraft noise on communities;
- A universally applicable, univariate logistic regression function is the best way to predict residential effects of aircraft noise exposure in all communities;
- Airports are undoubtedly the best and highest economic uses of land;
- Protection of public investment in airports is of paramount regulatory importance;
- Aircraft noise exposure becomes excessive only at sound levels greater than L_{dn} = 65 dB[7];
- No explicit rationale is required to support the definition of $L_{dn} > 65$ dB as the start of "too much noise";
- FICON (1992) is the sole source of technical authority, and FAA is the sole source of regulatory authority, in such matters; and
- Residential acoustic insulation, private property purchases, land use planning, and operational changes, all as defined and strictly controlled by FAA, remain the only practical remedies for excessive aircraft noise exposure.[8]

As might have been expected, the acoustical engineers who first sought answers to the perennial questions found them within the realm of acoustical engineering. Experience has shown, however, that these questions are not exclusively, nor even primarily, answerable by purely acoustic reasoning, nor by generic, correlation-based statistical methods. Pragmatic answers to these traditional questions inevitably require non-technical value judgments, as well as assessments of their political and economic consequences. Given the demonstrated inability of aviation noise regulatory policy to achieve its own goals, better answers also require innovative analytic approaches. The succeeding sections lay out the context in which technical, legislative, judicial, urban planning, commercial, economic, and other societal interests in aircraft noise regulation have played out in the United States.

1.4 History of Technical Concern with Aircraft Noise Management

The earliest concerns with the then-unprecedentedly high continuous sound levels created by aircraft jet engines (BENOX 1953) focused on individual effects in military occupational settings, with only minor attention to community noise implications (Borsky 1952.) Hearing damage risk and speech interference were the major concerns, but community noise-induced (and vibration-induced) task interference and rocket launch noise were also recognized as related issues. Research associated with such concerns persisted through the 1960s. Much of this early research was conducted by NASA and the U.S. Air Force Aeromedical Research Laboratory at Wright-Patterson Air Force Base.[9] Research in more recent decades has focused on population-level (rather than individual-level) effects of aircraft noise exposure.

Prior to the start of scheduled passenger jet service in 1958 at New York's Idlewild (now JFK) airport, adverse public reaction to aircraft noise nationwide was regarded as a minor, localized nuisance. Concerns with non-occupational exposure to aircraft noise were limited primarily to military base housing; to communities near military airfields; and to neighborhoods near runway ends at large civil airports. Even after the start of large jet transport operations at major urban airports, another decade passed before jet aircraft operations became commonplace at smaller regional airports. Today, community opposition to aircraft noise exposure is more the rule than the exception at airports worldwide, particularly as efforts are made to increase airport capacity or change operational patterns.

1.5 Basics of Contemporary Approach to Aircraft Noise Regulation

Regulation of aircraft noise in the United States is aimed primarily at (1) controlling noise emissions of individual aircraft types (within weight classes) offered for sale, and (2) managing the long-term, cumulative noise exposure that airport operations produce in nearby neighborhoods. Limits on noise emissions of individual aircraft offered for sale in the United States are coordinated with those of the United Nations' International Civil Aviation Organization (ICAO), while limits on fleet noise emissions in airport environs vary among member nations.

Regulation of aircraft noise effects is blind for most purposes to distinctions between indoor and outdoor exposure of household residents, even though the two forms of exposure may differ by ~20 or more dB.[10] The long-established practice of basing regulatory policy on outdoor residential rather than indoor personal noise exposure is necessary because the total personal exposure of individuals to aircraft noise cannot be easily or reliably estimated (Fidell 2015). Geographic associations between outdoor residential noise exposure and annoyance prevalence rates are therefore completely insensitive to the personal aircraft noise exposure of individual neighborhood residents. Although residents of airport-vicinity neighborhoods may spend very different amounts of time at home, and their waking hours may be spent in noise environments very different from those of their homes, their aircraft noise exposure for regulatory policy analyses is that outside their residences.

Further, policy threshold levels are defined for most purposes (e.g., to identify categories of land use compatibility) at 5 dB exposure intervals. Greater precision of definition of policy points is considered unwarranted, if not unachievable. In other words, regulation of aircraft noise exposure in airport environs is based on a step function-like association between cumulative outdoor noise exposure levels of households in areas within specified noise exposure boundaries, not on the actual personal aircraft noise exposure of individuals.[11] As discussed at greater length in Sect. 3.9.5, geographic associations of this sort, routinely inferred from cross-sectional field studies, are the most abundant form of epidemiological evidence about aircraft noise effects on people. They are also the least costly and weakest.[12]

Chapter 2
Chronology of U.S. Aircraft Noise Regulation

2.1 Landmarks in the Development of U.S. Aviation Noise Regulation

The following sub-sections describe some of the major legislative and regulatory actions affecting aviation noise in the United States. See Bennett (1982) for discussion of additional case law, and Sabel (2004) for further discussion of early legislation affecting air commerce.

2.1.1 Pre-World War II Years

Formal regulation of airports in the United States began with the Air Commerce Act of 1926 (Falzone 1999). The Act treated airports as individual entities under the "jurisdiction and control of municipalities concerned", rather than as components of an as-yet non-existent nationwide air transportation system. The 1926 Act also asserted that the U.S. government "possessed" all airspace, a claim invalidated by the Supreme Court 20 years later in *Causby v. U.S.* (328 U.S. 256, 1946) (*v.i.*).

The Civil Aeronautics Authority ("CAA") Act of 1938, which established the agency later re-named the Civil Aeronautics Board ("CAB"), reaffirmed local control of airports, even while setting the pattern for decades of federal regulation of passenger aviation.[13]

2.1.2 Post-World War II Years

As late as the mid-1950s, commercial aircraft operations were so sparse, and propeller-driven commercial aircraft sufficiently small and (relatively) quiet, that civil aircraft noise could plausibly be viewed as a problem confined to neighborhoods

S. Fidell and V. Mestre, *A Guide To U.S. Aircraft Noise Regulatory Policy*,
https://doi.org/10.1007/978-3-030-39908-5_2

adjacent to runway ends.[14] Modern U.S. positions on the definition of "significant"[15] noise exposure, and on disclosure of adverse environmental impacts of aircraft noise, can be traced to the start of jet-powered military aviation in the late 1940s and early 1950s. These origins antedate by a decade the 1958 Federal Aviation Act which created FAA.

High levels of aircraft noise exposure were viewed primarily as an occupational hazard at the start of the jet era, and residential concerns with aircraft noise exposure were confined mostly to military base housing. Many U.S. civil airports in the early post-World War II era were typically utilitarian, lightly-used former military facilities. Beginning in the 1960s, renovations and upgrades of runways and landside facilities that accompanied the expansion of flight operations and the enormous growth of in-terminal foot traffic began to transform large airports into shopping mall-like facilities with runways nearby.

Following a 1956 mid-air collision of a DC-7 and a Super Constellation over the Grand Canyon, Congress in 1958 assigned the Federal Aviation Agency (later re-named the Federal Aviation Administration), a dual mandate with regard to regulation. The 85th Congress described Public Law 85-726 as an Act "…to create a Federal Aviation Agency, to provide for the regulation *and promotion* [*emphasis added*] of civil aviation in such manner as to best foster its development and safety…". In other words, Congress directed the new agency to regulate for two purposes: first, to provide safe and efficient air transportation services for the nation, and second, to promote civil aviation by fostering its development. The original act makes no mention of minimizing adverse environmental impacts of aviation or protecting communities from them.

As interpreted by FAA, the agency's dual mandate required that policy positions on permissible levels of public exposure to aircraft noise *not* infringe on promotion and development of civil aviation. FAA has never had a direct mandate to promote municipal development, nor to vigorously protect airport neighborhoods from excessive aircraft noise exposure.[16] On the contrary, FAA's primary noise regulatory goal for decades was to protect public investment in airport infrastructure, or in essence, to protect airports from communities.[17] In practice, this goal meant that FAA's aircraft noise regulatory positions (e.g., on the definition and disclosure of "significant" noise exposure; on eligibility for mitigation of noise impacts; and on land uses considered "compatible" with operation and expansion of airports) were intended primarily to support the interests of airports, air carriers, express shippers, airframe and engine manufacturers, and air travelers.

2.1.3 *Influence of United States v. Causby and Griggs v. Allegheny County Supreme Court Decisions*

During World War II, the U.S. military flew aircraft at altitudes as low as 83 feet above the Greensboro, NC farm of Thomas Causby, panicking his chickens. The

U.S. government cited its claim in the 1926 Air Commerce Act that the government "possessed" all airspace as supporting its right to conduct such low altitude overflights, but lost in lower courts to Causby's argument that the government had taken the value of his land in violation of the "takings" clause of the 5th Amendment. The ruling against the United States in *United States v. Causby* (328 U.S. 256, 1946) established a legal precedent that a demonstrable economic loss due to aircraft noise exposure was required to prevail in such "inverse condemnation" litigation.

Sixteen years later, the Supreme Court built on this precedent in *Griggs v. Allegheny County* (369 U.S. 84, 1962). The decision on aircraft noise liability, reached only four years after the start of commercial jet service in the United States, set the pattern for much subsequent aircraft noise regulatory policy. On land that Allegheny County had acquired for airport use, the proprietor of Pittsburgh's airport (PIT) had constructed a runway that ended about 1 km from plaintiff's property. A Pennsylvania Court of Common Pleas awarded plaintiff $12,690 for a 5th Amendment "taking" of an air easement by the County that caused airplanes to fly at low altitudes over plaintiff's home, thereby interfering with the plaintiff's beneficial use and enjoyment of his property.

The County appealed the trial court's decision to the state Supreme Court, which found that the County was *not* liable for any such taking. The Pennsylvania Supreme Court's reasoning was that the County had fully complied with the provisions of the U.S. National Airport Plan (49 U.S.C. 1101 *et seq.*), and that the Federal Government had provided much of the funding for runway construction. The plaintiff appealed the reversal of the trial court's decision to the U.S. Supreme Court.

The Supreme Court could in principle have assigned partial to full liability for noise damages to any combination of three parties:

The airlines, which owned or leased the noise sources that interfered with plaintiff's beneficial use and enjoyment of property;
The Federal Government, which had provided the County with much of the funding for runway construction and controlled the movements of the airlines' aircraft in flight; or
The airport proprietor, which had made all decisions about runway siting, and leased gates to airlines for their flight operations.

The U.S. Supreme Court reversed the decision of the Pennsylvania Supreme Court, faulting the airport proprietor alone for not acquiring sufficient property to build the runway. As a consequence, "ownership" of liability for aircraft noise damages in the United States has ever since rested with airports, with no direct financial consequence for airlines or FAA. Had the Court decided otherwise in Griggs versus Allegheny County, U.S. aircraft noise regulatory policy would likely have developed in very different ways.

For example, if the Federal Government had been held liable in full or in part for aircraft noise damages, airport layout plans would have received much closer scrutiny, and far more critical analyses in FAR Part 150 studies, Master Planning exercises, and proposals for construction of airport infrastructure. If the airlines had become the "owners" of noise liabilities, greater pressures for reduction of aircraft engine noise would have been applied sooner; the costs of operating marginally profitable flights might have tempered enthusiasm for excessive hub-and-spoke network development; development of modern performance-based navigation ("PBN") flight procedures might place greater emphasis on ground-level noise exposure than on fuel and flight crew cost savings; and so forth.

Had the Supreme Court found some combination of airlines, airport proprietors, and the Federal Government liable, all of the liable parties would have been incentivized to collaboratively minimize and mitigate aircraft noise impacts in airport neighborhoods. Without feared noise liabilities, commercial proprietors (or even airlines) might have contested local government ownership of airports. As is, FAA plays little substantive role beyond financial planning support and *pro forma* approval of airport layout plans in local and regional development of airport infrastructure, and individual airlines and air cargo carriers are effectively invisible in most airport/community controversies. If liability for noise damages had not been assigned exclusively to airports, FAA might plan airspace use with lesser concern for minimizing airline flight time costs than for residential aircraft noise exposure. Further, airports might have access to additional means and greater flexibility in crafting compromises to resolve noise-related confrontations with communities.

Although the Griggs decision was feared at the time to have opened the door to widespread litigation against airports for noise damages, such has not been the case. Property owners typically encounter many difficulties in proving aircraft noise damages. Since damages are nearly always expressed as diminutions in property values, valuing lands near airports usually involves courtroom battles of assessors. When valuing property, plaintiffs stress adverse effects of aircraft noise, air pollution, and traffic congestion on property values, while defendants stress proximity to jobs, travel, and resources, and a host of local factors ranging from proximity to schools, convenient access to shopping, water views, and similar neighborhood amenities. (The larger issues of disproportionate sharing of regional economic benefits and local impacts of noise generated by a large airport are noted in Sect. 3.10.)

2.1.4 Slow Recognition of Need for National Aircraft Noise Regulation

Harper (1988) faults the FAA for its inattention during its first decade to aircraft noise as a nationwide problem. Smith (1989) likewise points out that governments, in their haste to facilitate the growth of air commerce in the early days of jet transport aircraft, bear some of the responsibility for exacerbating noise problems. Since the Supreme Court had immunized FAA against liability for aircraft noise damages (per Sect. 2.1.3), and the agency's charter required it to promote civil aviation, FAA's preference for the two decades following the decision was for a *laissez-faire* approach to aircraft noise regulation. This preference was based on FAA's apparent belief that stricter regulation of aircraft noise would invite industry criticism of inconsistency with FAA's original charter to promote civil aviation. Aviation-related industries (airframe and engine manufacturers, airlines, airport operators and their trade associations) also pressed for protection against assumption of any liability for aircraft noise damages. FAA responded to these pressures by attempting to facilitate voluntary cooperation on aircraft noise problems among aircraft and engine manufacturers, airlines, and airport proprietors.

As discussed further in Sect. 5.2, the United Nations' International Civil Aviation Organization ("ICAO") has long advocated a "balanced" approach to minimizing aircraft noise effects—that is, a combination of source-level reduction, airport-compatible land use planning, and operational controls to reduce noise impacts.[18] ICAO's balanced approach strategy has been largely ineffective for land use planning purposes in the United States. The failures of ICAO's balanced approach in the United States are due to several factors: federal pre-emption of local discretion in operational controls; differences in State land use legislation nationwide; a lack of Federal authority over local land use planning; airport proprietors' lack of authority over uses of land adjacent to airports that they do not own; and landowners rights under the Fifth ("just compensation") Amendment to the U.S. Constitution.

2.1.5 Continued Industry Concerns with Liability and Operational Restrictions

The potential liability (for airports) imposed by a noise policy that sets limits for noise-sensitive land uses and for development of compatible land use plans has been a central concern of aircraft noise regulation from its very beginning. The concern for liability for damages caused by aircraft noise exposure has strongly favored creation of large buffer areas around airports.[19] This concern creates a paradoxical policy inconsistency, however: airports must explain to residents of nearby neighborhoods that their aircraft noise exposures are insufficient for mitigation or compensation (i.e., are at levels lower than $L_{dn} = 65$ dB), while simultaneously insisting to owners of undeveloped land of similar noise exposure that their property is too noisy ($L_{dn} >$ 60 dB) for conventional residential development.

Further, the emphasis of noise regulatory policy historically has been on near-term concerns for liability around long-established urban airports. Residential land use patterns near such airports were often established before the introduction of jet aircraft. Prevention of longer term residential encroachment at less-developed airports appears to have been of lesser concern. Prevention of encroachment of airport noise on communities (as by growth and changes in numbers, types, and locations of flight operations, changes in fleet composition, and construction of additional airport infrastructure) seems to have been of yet lesser concern.

FAR Part 150 appears to have been intended primarily to discourage community encroachment around airports in less-developed areas, such as reliever and regional airports. Given that the "Noise Control Program" phase of a Part 150 study still maintains a list of alternatives that are effectively prohibited under the 1990 ANCA, management of existing airport-vicinity noise problems at larger airports was at best an afterthought.

These emphases were arguably reasonable in the early days of policy development, but are less so today, as aviation has largely avoided liability for noise damages to people living within areas that exceed FAA's $L_{dn} = 65$ dB guideline for residential land uses. Approximately 400,000 people reside within the $L_{dn} = 65$ dB contour today (FAA 2018). Residential sound insulation programs cover a portion of these homes, but are by no means fully inclusive. Not all of the 400,000 people within the $L_{dn} = 65$ dB contour would qualify for sound insulation under current guidelines. Most residential sound insulation programs are focused on single family detached dwellings, and usually do not extend to apartments or condominiums.[20]

Likewise, homes built after 1998 are assumed to have been designed to protect interior areas from aircraft noise exposure, and do not qualify for residential sound insulation programs. Little data supports this assumption in most states. Further, a home qualifies for federally sponsored sound insulation only if its interior noise levels can be shown empirically to exceed $L_{dn} = 45$ dB. The criterion of an interior DNL of 45 dB requires that the test be done with windows closed, regardless of whether a home has adequate ventilation to meet building code fresh air ventilation requirements.

Airport proprietors have long been almost exclusively municipalities or other regional government entities. With a few notable exceptions (e.g., the Port of New York Authority – now the Port Authority of New York and New Jersey, as described in a first-person account in Chap. 6 of Beranek 2008), airports have historically lacked both influence and authority to persuade, much less compel, industry to develop and operate quieter jet transports. Unless compelled by inverse condemnation, nuisance, or personal injury lawsuits, individual airports were reluctant through the early 1980s to adopt measures to meaningfully reduce aircraft noise exposure (and hence, adverse reactions to aircraft noise) in airport communities.[21]

Smith (1989, pp. 22–23) reproduces a portion of a 1966 letter from FAA to U.S. aircraft manufacturers warning them that "… it could well be that the most significant deterrent to continued growth *[of commercial aviation]* is related to problems associated with aircraft noise". The letter further warned that FAA might be forced by new legislation to "… require compliance with noise standards as well as compliance with safety standards as a condition to the issuance of future type certifications"; and that the agency was concerned about "…the engulfing of lands immediately surrounding our airports by urban communities…".[22]

The Housing and Urban Development Act of 1965 (Public Law 89-117) included a provision to study the effects of aviation noise on "economic loss and hardship suffered by homeowners as the result of the depreciation in the value of their properties…".[23] This legislation was intended to protect federally backed mortgages introduced by this landmark legislation. At the time of this writing, half a century later, the mandated report to the President has not been published.

2.1.6 Genesis of FAR Part 36

Faced with nationwide political dissatisfaction in the late 1960s with growing adverse community reaction to noise in airport environs, and fueled by substantial increases in numbers of jet aircraft operations, Congress amended the FAA's charter in 1968 via an act "to require aircraft noise abatement regulation and for other purposes" (Public Law 90-411, 49 U.S.C. Section 1431). The amendment forced FAA for the first time to treat aircraft noise as a national, rather than a local issue.

FAA's initial response to its new direction from Congress was issuance in 1969 of Part 36 of the Federal Aviation Regulations, describing noise certification requirements for source control of aircraft noise, as discussed in greater detail in Sect. 4. The regulation established noise limits by weight categories[24] for new aircraft offered for sale in the United States. Setting noise limits for new aircraft types had little immediate effect on the composition of the existing fleet, however.

2.1.7 National Environmental Policy Act of 1970 ("NEPA")

Passage of Public Law 91-190, the National Environmental Policy Act of 1970, focused regulatory attention on prediction and disclosure of noise impacts associated with airport infrastructure projects. The act requires agencies proposing major federal actions to fully disclose to decision makers the effects of such actions on the environment. More than any prior legislative action, NEPA necessitated formalization of FAA's methods of prediction and its interpretive policies regarding actual and prospective aircraft noise effects ("impacts", in environmental impact assessment jargon) on residential populations around airports.

The Council on Environmental Quality, a body created by NEPA, required FAA to spell out its procedures for evaluating the significance of aircraft noise impacts. FAA's response was Order 1050.1A, since amended five times at the time of this writing. The Order specifies "policies and procedures to ensure agency compliance with the National Environmental Policy Act", as well as additional requirements imposed by the Council on Environmental Quality and Department of Transportation (DOT) Order 5610.1C, "Procedures for Considering Environmental Impacts". NEPA is a procedural statute, however, rather than a substantive one. It requires FAA to take certain defined steps to assess the environmental impacts of a project, but does not in itself require the agency to meet any standard, nor to mitigate any noise impact that it discloses.

2.1.8 FAA's Definition of the Significance of Noise Impacts

Among its numerous provisions, FAA Order 1050.1 formalizes a set of cumulative noise exposure thresholds for disclosing and mitigating changes in aircraft noise levels, and for defining the "significance" of aircraft noise exposure. The significance of noise exposure is most reasonably defined in terms of its effects on individuals and communities, and was in fact so defined in the original Community Noise Rating scheme (Rosenblith et al. 1953). Rosenblith et al. associated community response to aircraft noise exposure at a CNR value less than 100 (subsequently transformed into NEF units, and thence into units of DNL as $L_{dn} = 65$ dB), with "essentially no complaints", even though such exposure might "interfere occasionally with certain activities of the residents" of communities with such exposure.

As noted in Sect. 3.1, FAA adopts the FICON (1992) position that "…the percent of the exposed population expected to be highly annoyed (%HA) *[is]* the most useful metric for characterizing or assessing noise impact on people". One might therefore logically expect FAA to define the significance of aircraft noise exposure directly in terms of a percentage of the population describing itself as highly annoyed. FAA Order 1050.1, however, offers no such straightforward definition of the significance of noise impacts. Instead, it expresses the significance of noise exposure solely in terms of noise exposure itself, in units of decibels, not in units of noise effects.

FAA abandoned the original definition of aircraft noise impact (in terms of complaints) after the agency adopted DNL as its preferred measure of noise exposure as required by ASNA. The 1992 FICON curve (see Sect. 3.2.4) provides the only linkage between noise exposure levels and the noise effect that FAA considers the most appropriate measure of noise impacts – the prevalence of noise-induced high annoyance in a community. Since the prevalence of high annoyance that the FICON curve associates with (all) transportation noise exposure is 12.29% at $L_{dn} = 65$ dB, it is difficult to escape the conclusion that FAA considers that exposure of more than 12.29% of the residential population to highly annoying aircraft noise constitutes a "significant" noise impact.

The arbitrariness of considering 12.29% of the population as an indicator of the presence of significant aircraft noise impact is likewise difficult to escape. The definition begs a host of questions such as "What technical rationale supports consideration of a prevalence rate of high annoyance of 12.29% in a community as a threshold of significant noise impact?"; that is, "Why 12.29%, rather than 10 or 15%?"; "Why is there no rigorous, quantitative cost-benefit analysis to support FAA's regulatory policy thresholds?"; "Why define a single, constant noise exposure level as significant in all communities when it is well known that communities with similar aircraft noise exposure differ greatly from one another in annoyance prevalence rates?" For that matter, "Why define the significance of noise impacts in units (i.e., DNL) that do not map uniquely into impact units (i.e., prevalence of high annoyance) in different communities?"

In lieu of any other definition of significance, FAA's retention of a definition traceable to CNR = 100 serves as a de facto definition of a prevalence rate of high annoyance of 12.29% as a threshold of significance of noise impact.[25] Current policies fail to address the large differences in annoyance prevalence rates at the same noise exposure levels in different communities in several ways:

(1) A nationwide definition of the significance of aircraft noise exposure is often offered as a necessity to avoid the airline-dreaded "patchwork quilt" of regulation at different airports. Until FAA lost its charter to promote civil aviation in 1996, allowing different airports to establish different cumulative exposure thresholds of significant noise exposure would almost certainly have been challenged by aviation industry interests as inconsistent with FAA's charter to promote civil aviation.[26]
(2) Current policy tacitly assumes that the 14th Amendment to the U.S. Constitution principle of equal application of regulation of law requires a universal definition of significance of noise exposure in all communities. This is a somewhat simplistic position that is undermined by the fact that defining a unique sound-level threshold for noise exposure in all communities does not insure equal *effect* of regulation in many of them. (The FICON logistic regression curve greatly under- and over-predicts annoyance prevalence rates in any but a hypothetical, perfectly average community.) Further, the annual average day conditions for which DNL values are calculated may never be experienced on any actual days, compounding efforts to explain the nature of noise exposure modeling to the public.
Both the responsibility and authority for land use controls are local, not federal, in any event. This leaves unanswered the question of why the aviation industry has long been resistant to a federal policy level lower than $L_{dn} = 65$ dB.
(3) Current regulatory policy continues to ignore the role played by non-acoustic factors[27] in determining the effects of noise exposure on communities. This position is a vestige of 1950s-era efforts to predict community response to noise exposure exclusively via measurement or modeling of noise exposure levels. The limitations of this position are now well understood, and formal, systematic analytic means (*cf.* Fidell et al. 2014) now exist for quantifying the net influence of non-acoustic factors on empirically observed annoyance prevalence rates.

2.1.9 Noise Control Act of 1972

The Noise Control Act of 1972 (Public Law 92-574) established as a matter of national policy that Americans should live free from noise that jeopardizes their health and welfare. The act was not concerned specifically with aircraft noise, nor even with transportation noise alone. Although the act was not concerned only with aircraft noise, it required FAA to develop rules and regulations regarding aircraft noise. Among the Act's other provisions, the Environmental Protection Agency ("EPA") was empowered to coordinate federal research and activities in noise control. From

1972 to 1980, this created bureaucratic tension between EPA's Office of Noise Abatement and Control ("EPA/ONAC") and FAA's Office of Environment and Energy ("FAA/AEE"). The clash between EPA/ONAC and FAA/AEE echoed their agencies' respective charters: EPA/ONAC was established to protect public health and welfare, whereas FAA's charter included promotion of civil aviation.

A 1973 EPA "Aircraft-Airport Noise Report to Congress" found that new regulations for flight and operational noise controls would be needed, including takeoff, approach and landing procedures, and minimum flight altitudes. Further, EPA called for amendments to FAR Part 36, as well as regulations to control and reduce noise emissions from existing aircraft. From the aviation industry's perspective, EPA/ONAC's establishment infringed on FAA's former sole authority for regulating aircraft in a manner consistent with promotion of the industry.

Another Report to Congress mandated by the 1972 Noise Control Act (von Gierke 1973) evaluated noise metrics and criteria for "acceptable" aircraft noise levels. von Gierke (correctly) recognized that the controversial issue of aircraft noise impact analysis was not properly a purely technical issue, but "a social, ethical, and economic one: what percentage of the population shall be protected and at what price?" This depth of understanding has been absent from much subsequent policy debate.

The best-known report produced by EPA/ONAC was its 1974 "Levels Document", whose full title ("Information on Levels of Environmental Noise Requisite to Protect Public Health and Welfare with an Adequate Margin of Safety") reiterates the legislative language mandating the report's production. The Levels Document introduced Day–Night Average Sound Level as the primary measure of environmental noise exposure, and identified an exterior level of $L_{dn} = 55$ dB (i.e., 60 dB, with a 5 dB margin of safety), and an indoor level of $L_{dn} = 45$ dB, as levels of environmental noise that posed no threats to health and welfare. The document took pains to emphasize that the noise exposure levels that it identified made no provision for economic or practical feasibility.

The tension between EPA/ONAC and FAA/AEE continued until the 1981 budget for EPA/ONAC was reduced to zero by the Reagan administration, without rescinding Public Law 92-574. This action created a paradox in which EPA's statutory obligation to abate noise remains intact to this day, but the agency has been deprived of resources for carrying out its obligation. One practical consequence of the termination of EPA's ability to act under the Noise Control Act was not merely that FAA regained exclusive custody of aircraft noise regulation, but also that state and local noise control efforts became pre-empted and de-funded (Shapiro 1992). Subsequent legislation (see below) further preserved FAA's sole prerogative to regulate aircraft noise without explicit responsibility for protection of public health and welfare, until the 1996 Department of Transportation Re-Authorization eventually rescinded FAA's charter to promote civil aviation.

2.1.10 Aviation Safety and Noise Abatement Act of 1979 ("ASNA")

Congress enacted the Aviation Safety and Noise Abatement Act of 1979 ("ASNA") to direct FAA "to establish a single system of measuring noise and the impact of noise on individuals to be used to measure noise at airports and their surrounding areas and to establish land uses for such areas which are compatible with such noise levels". FAA eventually did so six years later by issuing its "Airport Noise Compatibility Planning" (14 CFR Part 150) regulation to implement ASNA.

The six-year delay from passage of ASNA in 1979 to issuance of FAA's implementing regulation (FAR Part 150) in 1985 was due in part to bureaucratic skirmishing between EPA's Office of Noise Abatement and Control and FAA's Office of Environment and Energy. As described above, the 1972 Noise Control Act had given EPA authority for regulatory oversight of transportation noise, and EPA's 1974 "Levels Document" had adopted a particular noise metric ("Day Night Average Sound Level", or DNL) as its preferred measure for characterizing community noise impacts. FAA resisted adoption of EPA's community noise metric—itself borrowed from California's aviation noise regulations – for years. FAA resisted acceptance of DNL as a measure of aircraft noise until forced by advances in understanding the predictability of aircraft noise annoyance (EPA 1973; Schultz 1978) to publish the much-delayed implementing regulation for ASNA.

2.1.11 Airport Noise and Capacity Act of 1990 ("ANCA")

Several U.S. airports adopted noise restrictions in the mid-1980s that had the potential to constrain operations. These included John Wayne Airport in Orange County, Long Beach Airport, and Santa Monica Airport (all in California) and Denver's new airport in Colorado. A brief simple description of their programs is as follows:

- John Wayne Airport: The Phase II Access Plan was the result of a stipulated settlement agreement signed in 1985. The access plan limits the number of commercial operations by noise class, establishes daytime and nighttime single-event noise limits at fixed noise monitor locations, sets an annual passenger limit, and establishes a night curfew for commercial operations and other controls on the development of the airport. It is often cited as the strictest noise control program in the United States.
- Long Beach Airport: Long Beach Airport's Noise Ordinance sets in place a nighttime curfew, single-event noise limits at fixed noise monitor locations, and an airport noise budget based on cumulative noise levels. It was also the result of a 1985 settlement agreement to ongoing litigation.
- Santa Monica Airport: Santa Monica Airport is general aviation airport with a long history of noise litigation. A 1984 Settlement Agreement confirmed a nighttime

curfew, single-event noise limits at fixed noise monitors, and limits on the aviation developments on the airport.

- Denver: In 1988, the City and County of Denver, Colorado, wishing to develop a new airport in an adjacent rural county to replace its existing Stapleton Airport, was compelled to assure residents of neighboring Adams County that aircraft noise from operations at a new airport would never exceed cumulative noise levels in Adams County produced by 1987 operations at Stapleton airport. Adams County insisted on a formal inter-governmental agreement that included substantial payments that Denver would owe Adams County should long-term, cumulative aircraft noise from the new airport exceed "noise exposure performance standards" at 101 points on the ground in Adams County. In return for this binding contractual agreement, Adams County consented to annexation by Denver of about 50 square miles for construction of a new airport.[28] At the time of this writing, DEN has been ordered to pay a cumulative total of approximately $75 million to satisfy its obligations under the inter-governmental agreement.

The implications of such agreements in the face of local opposition to such projects were of particular concern to the aviation industry. So too were the possibilities of exhausting runway capacity at busy airports,[29] and operational restrictions on airport use, such as night curfews and caps on numbers of permissible flights. By the late 1980s, aviation industry interests were alarmed by the possibility of proliferation of such local government agreements beyond California and Colorado.

FAA, whose charter at the time included promotion of civil aviation, was not a signatory to the agreement between the two local government agencies in Colorado, and had no authority to prevent additional agreements of the same sort. As described by Suter (1991), industry lobbied vigorously for legislation to drastically increase federal control over airports, in order to effectively preclude airports and local governments from entering into future agreements limiting airport operations in any manner. ANCA, attached to a must-pass bill and hastily passed without benefit of public hearings in the heat of a budget crisis, was the result.

The major provisions of ANCA include (1) intentionally difficult, complex, and costly procedures in FAR Part 161 for seeking FAA approval of any compulsory operational changes (night curfews, caps on maximum noise levels, numbers of operations, noise-based landing fees, and the like); and (2) de facto veto power for airport users over proposed operational restrictions. ANCA also granted airports the authority to assess head taxes ("passenger facility charges"), while denying the same authority to local governments. As an ostensible concession to airport communities, the noisiest ("Stage 2") commercial aircraft then flying were to be withdrawn from service a decade after passage of the act.[30]

2.2 Revocation of FAA Charter to Promote Civil Aviation

For decades, Congress viewed FAA's maintenance of a light regulatory hand to avoid inhibiting the growth of commercial aviation as a proper interpretation of the dual mandate assigned to the agency in 1958. Congress eventually came to view the dual mandate as an inherent conflict of interest, however, and rescinded FAA's dual mandate in the 1996 Department of Transportation Reauthorization Act. Public Law 104–264 relieved FAA of responsibility for promoting civil aviation, limiting the agency's regulatory remit to maintenance of a safe and efficient national air transportation system.

Despite FAA's loss of its charter to promote civil aviation, the agency's pre-1996 policies on permissible levels of public exposure to aviation noise remain in effect, without revision. These policies effectively subordinate community to aviation industry interests, but may soon be revisited, as noted in the next sub-section.

2.2.1 *FAA Reauthorization Act of 2018*

The 115th Congress included several aircraft noise-related provisions in Public Law 115-254 (the FAA Reauthorization Act) at the urging of the 40-member Congressional Quiet Skies Caucus. These include a requirement for FAA to finish within one year its long-delayed evaluation of alternatives to the Day–Night Average Sound Level (DNL) and 65 dB threshold of significance of noise effects. The Act also directs FAA to consider diverging departure flight paths or lateral spacing to address community noise concerns. The Act further required FAA to undertake studies about health effects of noise on local communities, and thereafter to make recommendations within two years for revising land use compatibility guidelines.

Some of these requirements could in principle compel FAA to reconsider and clearly state its rationale for NEPA-related definitions of significant aircraft noise impact, as well as its eligibility criteria for acquiring private property and for participation in aircraft noise mitigation programs. Such reviews and policy updates have been very slow to develop.[31] Airport-vicinity aircraft noise problems are sometimes re-cast as safety or efficiency issues. For example, the recent introduction of NextGen air traffic procedures were designed solely for efficiency, with no consideration of using these new procedures to mitigate noise (beyond the use of lower power approach profiles). Nor have any rigorous, quantitative cost–benefit analysis supporting FAA's historical noise policy preferences been conducted in NextGen exercises.

Others of the aircraft noise-related requirements of the 2018 Reauthorization Act are wide of the mark. For example, the requirement for FAA to evaluate alternatives to DNL has been studied many times, including an exhaustive review of the topic a few years ago (Mestre et al. 2011). It is doubtful for reasons described in detail in Mestre et al. that yet another such study would support any different conclusions. In any event, the greater problem is not finding a holy grail of noise metrics that accurately and precisely predicts the prevalence of annoyance with aircraft noise in all communities, but rather a more fundamental lack of systematic understanding of noise-induced annoyance, and of how to incorporate such complexity into noise policy.

2.3 Other Developments Influencing the Growth of Aircraft Noise Exposure

Quite a few other federal legislative acts, administrative actions, and policy positions, as well as technical standards, although not aimed at disclosing or assessing aircraft noise impacts, nor at directly regulating aircraft noise, have nonetheless had important indirect consequences for the expansion and management of aircraft noise exposure in airport environs. These include provision of subsidies to the air transportation industry; officially sanctioned "land use compatibility" guidelines; regulations governing federally sponsored airport and airways facilities and airport revenues; and eligibility criteria for aircraft noise impact mitigation measures. Several of these are noted in the following sub-sections.

2.3.1 Air Mail Acts of 1925, 1930, and 1934

Scheduled daytime air mail service in the United States began under military auspices in 1918. It expanded into nighttime commercial service, managed by the Post Office Department, throughout the 1920s. A series of lucrative subsidies for mail carriage, as well as construction of emergency landing fields, beacons, and airfield lighting, were implemented specifically to foster the growth of civil aviation. The Contract Air Mail Act of 1925, which expanded the number of airmail routes and carriers, quickly attracted financial interests seeking subsidies to form private airline companies.

A small number of cartel-like holding companies, which controlled both airframers and transport companies, operated most of the airmail routes by the end of 1920s. This arrangement culminated in a scandalous "Spoils Conference" in 1930, at which the Post Office colluded with the cartels to discourage bidding by smaller airlines on future air mail contracts. The Air Mail Act of 1934 was intended to rectify such abuses. It eventually led to the formalization of regulated air carrier operations that lasted until the 1978 deregulation of the air transport industry. Regulated air carrier operations for years guaranteed airline profits, which in turn led to an industrial oligopoly, economically unjustifiable numbers of low load-factor flights at major airports, and excessive arrival and departure noise.

2.3.2 *Airline Deregulation Act of 1978*

Public Law 95-504, although not specifically noise-related, nevertheless played a major role in sustaining the growth of aircraft noise exposure in airport environs. By abolishing CAB's role in approval of flight routes and by encouraging competition, it strongly supported continuing growth in annual numbers of aircraft operations. By replacing city-pair with hub-and-spoke flight routes, it concentrated the scheduling of operations at large hub airports. Morning and evening rush hours were replaced by as many as a dozen "banks" of flight operations at roughly hourly intervals throughout the day and evening. Deregulation also encouraged the growth of a new class of 50–70 passenger transports (regional jets, noisier than the short haul turboprops that they replaced, even if quieter than large jet transports) at many airports.

From a noise perspective, the net consequence of deregulation at hub airports was to periodically concentrate flight operations into recurring episodes of high runway demand throughout the day, thus putting increased pressure on peak airport runway acceptance rates. Airport managements at large hubs, eager to satisfy increased demand for air transportation services, responded by building new runways, terminals, gates, and complementary landside facilities. To residents of hub airport neighborhoods, the continual growth in operations ended mid-day relief from periods of heavy overflight noise.

2.3.3 *FAA Support for Airport Planning Exercises*

FAA controls access to funds needed to plan and construct airport infrastructure as a legacy of its former charter to foster the growth of civil aviation. FAA's primary funding source for supporting airport development is not its annual Congressional General Fund appropriation, but rather the Airport and Airway Trust Fund, established under the Airport and Airway Development and Revenue Act of 1970. Because the Trust Fund accumulates nationwide aviation-related excise taxes on passengers, cargo, and fuel, and distributions from the fund inevitably favor some airports over others (at least in the short term), it may be argued that users and stakeholders, not taxpayers, pay for such capital-intensive airport improvements. It is nonetheless airport-vicinity residents, rather than other stakeholders, who suffer increased exposure to aircraft noise at airports which receive subsidies that attract increased flight operations.

Individual airports' Master Planning exercises generally start with aviation activity forecasts[32] and analyses of current or anticipated future discrepancies between airfield demand and capacity. Airport Master Plans typically evaluate alternative development options, identify associated capital improvement needs, review the environmental consequences of alternative developments, and develop a preferred airport layout plan. As described by FAA Advisory Circular 150/5070-6B, U.S.

airports are eligible for substantial funding of the conduct of master planning exercises with twenty-year time horizons. (The long-term planning horizon gives airports a distinct advantage over community planning departments, whose interests do not often extend farther than the next election cycle.) Airport master plans may be updated as often as every five years to encourage airport-initiated development strategies – not to address regional air transportation needs, nor potential imbalances between airport desires for growth and local demand for air transportation services.

Funding the planning and implementation of airport growth is a legacy of FAA's original charter to foster the growth of civil aviation. Since the 1996 loss of FAA's charter to foster the growth of civil aeronautics, airport expansion projects can only be justified by arguing that they enhance the efficiency of the national air transportation system. Although some FAA construction grants include provisions to enhance the safety of ground operations, such grants may call into question the safety of existing airport designs.

From a community noise exposure perspective, nearly all major airport expansion projects (i.e., runway and terminal construction) facilitate growth in numbers of flight operations. Growth in air traffic may worsen aircraft noise exposure in already-affected neighborhoods, or newly expose other neighborhoods to aircraft noise. Grants that pay for major portions of airport infrastructure construction thus work at cross-purposes with the goal of reducing the numbers of airport neighborhood residents exposed to high levels of aircraft noise. As a rule, however, airline calculations of the profitability of adding flights at airports with lengthy flight delays do not take into consideration the costs of mitigating their noise impacts.[33]

Chapter 3
Aircraft Noise Effects on Individuals and Communities

Like other forms of transportation noise, aircraft noise annoys people, interferes with their speech, disturbs their sleep, and can interfere with performance of some non-verbal tasks. Aircraft noise does not pose a meaningful risk of hearing damage in residential settings (as it does in some occupational settings), but is often alleged to be linked with extra-auditory public health risks. For reasons discussed at length in Sect. 3.9, and despite many assertions to the contrary, the latter claims are only weakly supported by epidemiologic evidence, and remain controversial, even after decades of study.

3.1 Annoyance

Annoyance is by far the best documented and clearest effect of aircraft noise exposure on residential populations. As an attitude, annoyance is a covert and intangible mental process whose measurement requires self-report, most rigorously solicited as part of a carefully designed social survey.[34] FAA subscribes to the positions taken in Section 3.2.2.1, pp. 3–3 *et seq.* of the 1992 FICON report, that:

(1) "…The percent of the exposed population expected to be highly annoyed (%HA) *[is]* the most useful metric for characterizing or assessing noise impact on people", and
(2) "…The updated Schultz curve remains the best available source of empirical dosage- effect information to predict community response to noise".

FAA considers the prevalence of annoyance to be the "most useful" measure of community response to aircraft noise for two reasons: first, because annoyance is a stable attitude that is easily measured (simply by directly asking people); and second, because at least in principle, it is plausibly predictable from FAA's preferred (long-term, cumulative) measure of aircraft noise. The proper measure of "community response to aircraft noise" for FAA is simply the percentage of social survey

S. Fidell and V. Mestre, *A Guide To U.S. Aircraft Noise Regulatory Policy*,
https://doi.org/10.1007/978-3-030-39908-5_3

respondents who describe themselves as "very" or "extremely" annoyed when asked for their reactions to aircraft noise exposure over a recent week (or longer) time period.[35] Thus, in a pragmatic regulatory context, the term "community response" to aircraft noise implies nothing more than the prevalence of a consequential degree of aircraft-induced annoyance in a community.

Predicting the prevalence of high annoyance in communities as a mathematical transformation of DNL values ignores the influences of any and all non-acoustic factors on annoyance. Put another way, construing community response to aircraft noise exposure in terms of annoyance prevalence rates favors a high-level (long-term cumulative, or telescopic) view of the issue. (In contrast, quantifying community response in terms of speech interference would require a fine-grained, microscopic treatment of individual noise events.)

For many practical purposes, a telescopic view of community response to transportation noise is a more tractable and productive one than the microscopic alternative. For development and application of regulatory criteria to community reaction to noise, individual instances of noise intrusions into homes, as well as episodes of interference with speech and sleep, are treated as though they are subsumed in annoyance prevalence rates predicted by dose–response analysis.[36]

3.1.1 Consequences of an Acoustical Engineering Perspective on "Community Response"

As described in Sect. 3.2, the prevalence of noise-induced annoyance in a community can be reduced to a monotonic transform of aircraft noise exposure; and as described in Sect. 4.1.9, actual noise exposure itself can be replaced by a set of calculations and assumptions based on airport operational factors. Doing so completes the circle of reasoning supporting the regulatory treatment of community response to aircraft noise as a simple matter of acoustical engineering, with all of its attendant advantages and disadvantages.

The major advantage of this reductionist perspective is that it eliminates the need for regulatory consideration of the political and economic complexities of aircraft noise effects on actual communities. Engineering the "community" aspects out of "community response to aircraft noise", however, ignores the inconvenient variability of the responses of different communities to the same noise exposure.

Ignoring the fact that different communities respond differently to the same noise exposure guarantees that regulatory policy intended to protect people from exposure to highly annoying aircraft noise applies only to a mythical, completely average community, and that hypothetically predicted and actually observed "community response" may differ greatly. Ignoring such differences also implies acceptance of the following assumptions (among others):

- Community reaction to aircraft noise is caused exclusively by acoustic factors;
- All communities respond identically to aircraft noise;

- The only remedies for adverse community reactions to aircraft noise are acoustic;
- Regulation of aircraft noise exposure is a simple matter of picking threshold values of noise exposure that are tolerated in a hypothetically average community; and
- Computer-based modeling of aircraft noise exposure is essentially error-free.

Adopting such assumptions obviates needs to confront and solve the complexities of dealing with the actual interactions of airports and communities. For example, there is no need to inquire "What are the time constants of arousal and decay of community response to aircraft noise exposure?", because "community response", by definition, is a simple monotonic transform of an estimated, long-term average noise exposure value. Likewise, viewing community response simplistically—as no more than a mathematical transform of noise exposure—denies the possibility of reducing friction between airports and communities by non-acoustic means.

3.1.2 Difference Between Annoyance and Loudness

The attitude of annoyance differs most obviously from the sensation of loudness in that annoyance increases with the duration of a sound, whereas loudness does not increase with duration after about a quarter of a second—the so-called time constant of the ear (Garner 1947). The prevalence of a shared attitude of consequential annoyance in a community also differs from the individual behavior of making a noise complaint.

Aircraft noise regulation would be a simpler matter if duration were the only difference between loudness and annoyance. Sounds do not have to be loud or long in duration to be annoying, however. In some circumstances, sounds can be annoying even when only barely audible (Fidell et al. 1979). In residential settings, such sounds may include dripping faucets, footfalls and indistinct speech in adjacent apartments, and crying babies. The locus of control also affects annoyance—as when someone else's dog's barking is annoying (to you), but your own dog's barking is not (to you). The annoyance of sounds can also be situation-specific. In low ambient noise level outdoor recreational settings, noise from distant helicopters can be annoying when barely audible. A large audience attending an outdoor musical performance may be willing to pay for exposure to high-level sounds, while people in homes nearby may be annoyed by the same sounds at far lower levels.

Like other attitudes, annoyance with aircraft noise has multiple properties, including valence, extremity (intensity), accessibility, ambivalence, and embeddedness (Weiner et al. 2012.) Moreover, the attitude of annoyance has long been recognized (e.g., McKennell 1963; Borsky 1977; Job 1988; Fields 1993; Miedema and Vos 1999) to be influenced by many non-acoustic factors. These include other attitudes, emotions, situational variables, and personality traits, including fear, misfeasance, malfeasance, fairness, home ownership, introversion, and beliefs about the necessity and economic dependence on noise exposure.

For purposes of predicting community response to aircraft noise exposure, "noise sensitivity", other personality traits, and additional individual-level variables are of essentially no regulatory utility. In the first place, the incidence of individual traits that may pre-dispose or inure airport neighborhood populations to annoyance are unknowable in advance, and thus of meager use for predictive purposes. Even more relevantly, however, aircraft fly over all members of airport communities, whether they are highly sensitive or highly insensitive to noise exposure.

In short, exclusive reliance on a sound level value to predict the prevalence of annoyance in airport communities is a convenient, but rough and over-simplified expedient. Annoyance can be strongly affected by factors that sound level meters are simply not capable of measuring, nor noise level prediction software capable of modeling. (A modern approach to incorporating non-acoustic factors in analyses of annoyance prevalence data is described in Sect. 3.3.)

3.1.3 Prospective Disclosure of Noise Impacts for Airport Infrastructure Projects

For the sake of quantitative disclosure of environmental noise impacts, as required by NEPA, noise effects predictions are based on dose–effect relationships. Each such relationship embodies a tacit hypothesis: that whatever quantity is plotted on the abscissa causes variation in whatever quantity is plotted on the ordinate. The implicit hypothesis underlying prediction of the prevalence of annoyance in a community from its Day Night Average Sound Level is known as the "equal energy hypothesis". The hypothesis holds that the sound level, number, and duration of noise intrusions are equally influential as determinants of annoyance. In other words, the implicit hypothesis is that people integrate the acoustic energy of noise exposure exactly as do sound level meters.

This view is, of course, an over-simplification. Few truly believe that people perform continuous, true root-mean-square integration of acoustic energy over the course of a year on an average annual day basis, exactly as modern, true root-mean-square sound level meters and noise modeling programs can; nor for that matter, that annoyance is caused exclusively by acoustic variables. It has been appreciated at least since the first modern social survey of aircraft noise annoyance (McKennell 1963) that factors other than noise exposure affect annoyance judgments. Attempting to predict the prevalence of noise-induced annoyance in communities from DNL values alone is tantamount to asserting that peoples' reactions to aircraft noise are effectively determined by little sound level meters inside their heads.

3.1.4 DNL and the Equal Energy Hypothesis

DNL, a 24-hour frequency- and time-weighted average sound level, is the most common of the "equivalent energy" family of noise metrics. The family of noise metrics, originally defined in EPA's "Levels Document" (EPA 1974), also includes Sound Exposure Level (SEL) (for single events), and Equivalent Level (L_{eq}) (for arbitrary time periods). The equal energy hypothesis holds that annoyance with aircraft noise exposure is equally determined by the number, duration, and acoustic energy of aircraft noise exposure. In other words, the hypothesis asserts that it is not simply the maximum sound level of an aircraft overflight that controls its annoyance, nor the duration of individual overflights, nor the numbers of overflights, but the product of all three.

The empirical evidence in favor of the equal energy hypothesis is thin (e.g., Fields and Powell 1985), and counter-evidence has also been reported (Fidell and Pearsons 2003). The hypothesis is nonetheless so useful—albeit difficult to definitively confirm or reject by means of field experimental data – that it is broadly accepted as plausible. Doubts nonetheless persist about its applicability in the limit. Would people truly be indifferent between the annoyance of exposure to 500 and 1000 overflights per day if each of the latter overflights were 3 dB lower in sound level than each of the former? Would they also be indifferent between the annoyance of exposure to 500 and 1000 overflights per day if each of the latter overflights were of half the duration of the former? Field experiments of this sort are simple to imagine, but too impractical (expensive and time-consuming) to be conducted over a usefully wide range of exposure conditions.

Because DNL directly expresses the combined product all of the primary factors (level, duration, and number) which can reasonably affect the annoyance of aircraft noise exposure, it is equally sensitive to all of them. Thus, if the number of aircraft operations increases by a factor of two (i.e., by 3 dB), so does DNL. If the durations of aircraft overflights decrease by a factor of two, DNL also decreases by 3 dB. If the acoustic energy of individual aircraft overflights increases by 3 dB, so does DNL. In practical terms, this means that DNL is highly correlated with any sensible measure of aircraft noise.

The concept of an "equivalent, fully impacted population" is a corollary to the equal energy hypothesis. As described by von Gierke (in Harris 1979), the equivalent fully impacted population is estimated by adding together numbers of people exposed to varying degrees to environmental noise, and weighting numbers of people in 5 dB-wide exposure intervals. A total equivalent population fully impacted by a noise source may then be calculated by identifying and cumulating people exposed to partially annoying noises. Originally developed to satisfy NEPA-related disclosures of environmental impacts of proposed federal projects, the concept of an equivalent fully impacted population has fallen out of favor in recent years, in part because of the arbitrary nature of the weighting function (shown in Table 45.1 of von Gierke 1979) applied to exposure intervals.

The size of an equivalent, fully impacted population has also been criticized as an unhelpful measure of aircraft noise impact, because the quantity can be minimized not only by reducing aircraft noise exposure but also by reducing population sizes. In Pittsburgh, St. Louis, and Memphis, for example, thousands of homes have been purchased and demolished to make room for new runways. In Los Angeles and Seattle, residential property purchases have likewise eliminated neighborhoods antedating airports. Entire residential neighborhoods and even townships around some airports have disappeared, yielding equivalent fully impacted populations of zero.

3.2 Dose–Response Analysis

A dose–response function capable of informing aircraft noise regulatory analyses plots some measure of cumulative noise exposure on the abscissa (typically, DNL or DNEL), derived from either modeling or measurements against a measure of aircraft noise effect (typically, the prevalence of a consequential degree of aircraft noise annoyance in a community, as measured by social survey) on the ordinate.

3.2.1 *Early Dose–Response Analyses*

Figures 3.1 and 3.2, from EPA's 1974 "Levels Document", are early, somewhat schematic dose–response relationships between transportation noise exposure and the prevalence of a consequential degree of annoyance in communities. Figure D-16 of the Levels Document shows that "complaints and threats of legal action" were expected at cumulative exposure levels on the order of $L_{dn} = 65$ dB, even though fewer than 5% of community residents were believed to complain about aircraft noise at that

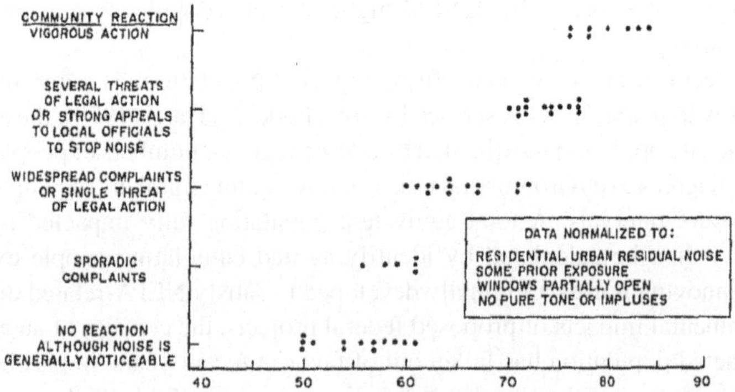

Fig. 3.1 Early relationship between noise exposure and its expected consequences

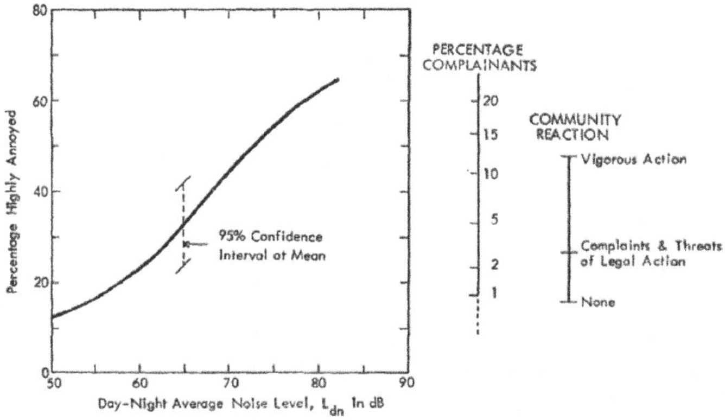

Fig. 3.2 An early dose–response relationship between DNL and the prevalence of annoyance

exposure level.[37] The information supporting the graphic was the best available at the time, but was based largely on a single study that conducted interviews in Boston, Chicago, Dallas, Denver, Los Angeles, Miami, and New York (Tracor 1971).

Initial versions of such relationships published in EPA's "Levels Document" (1974), based on modest amounts of information, suggested that about 30% of residential populations were highly annoyed by exposure to DNL values on the order of 65 dB.[38] As further described below, Schultz's (1978) early generic relationship for all forms of transportation noise (i.e., road and rail traffic as well as aircraft), associated exposure to a DNL value of 65 dB with a prevalence rate of high annoyance of 15.3%.

3.2.2 *The "Schultz Curve"*

The first modern and comprehensive synthesis of the literature (Schultz 1978) was not published until four years after publication of EPA's Levels Document. Schultz's synthesis was by far the most sophisticated of its time. It converted noise measurements made in different countries in different units into common units of Day–Night Average Sound Level, translated response scales and interpreted them in consistent terms, and established the convention of analyzing the prevalence of a consequential degree of annoyance, rather than average annoyance.[39]

Schultz's meta-analysis was very different from the conventional social survey research on community reaction to aircraft noise during the 1960s and 1970s, and initially highly controversial. Most early research analyzed the findings of individual studies, most of which characterized noise exposure in idiosyncratic units. Some objected to Schultz's conclusions on the grounds that characterizing community response in terms of "high" (rather than average) annoyance was statistically inefficient, in that the percentage of "high" annoyance responses did not reflect each individual response in the distribution of annoyance responses. Others objected to details of Schultz's conversion of noise metrics into common DNL units. Yet others were concerned that Schultz's meta-analysis was conducted at the level of community, rather than individual, reaction. Boeing (1973) had earlier objected to DNL on the grounds that "Community surveys have shown that noise alone is a poor indicator of annoyance. At present no subjective scale, including the new Ldn (*sic*) unit used in this report, can provide more than crude estimate of community response to a complex sound. Such deficient scales are not suitable for making major decisions and could result in costly mistakes". (Boeing's characterization of DNL as a "subjective scale" reveals a basic confusion about the difference between noise exposure measurements and interpretive criteria applied to such scales.)

Nonetheless, by the mid-1980s, Schultz's meta-analytic approach had become the conventional wisdom. For NEPA-mandated purposes such as disclosing the effects of proposed airport infrastructure projects, and for mitigating the effects of aircraft noise exposure on residential populations, noise exposure per se was widely recognized as an inappropriate yardstick for gauging the severity of aircraft noise impacts. It is not the noise created by aircraft itself, but the prevalence of high annoyance in a community engendered by noise to which the community is exposed, that is the proper measure of community response to aircraft noise. Although Schultz's synthesis was a landmark in transportation noise research, more recent research shows that aggregating road, rail, and aircraft noise into a single dose–response curve obscures the fact that people are a good deal more annoyed by aircraft noise than by road and rail noise.

3.2.3 FICON Position on Dose–Response Analysis

FICON (1992) generically recommends "dose–response" analysis as its preferred basis for environmental noise regulatory analyses. The recommendation is a sententious one, offered for little practical reason other than to bolster assertions that FICON's definitions of significant noise exposure and land use compatibility recommendations are "fact- and science-based". For FICON's recommendation to withstand close examination, it needs to be accompanied by many systematically supported specifics: the confidence that can be placed in summary dose–response functions in the face of enormous variability in annoyance prevalence rates in different communities with similar aircraft noise exposure; why aircraft noise regulatory policy thresholds are manifestly *not* based on formal dose–response analysis; the preferred form(s) of measurement and analysis of both dose and response; the influence of non-acoustic variables on annoyance prevalence rates; how regulatory criteria can emerge from meta-analyses absent political and economic criteria; and so on.

The FICON report is conspicuously silent on any systematic rationale for linking a given point on a dose–response relationship with a regulatory exposure threshold. This silence is a consequence of the fact that no dose–response relationship other than a step function is self-interpreting. All other dose–response functions, such as the ogival one favored by FICON (1992), require value judgments to support interpretive criteria. The $L_{dn} = 65$ dB criterion for the significance of aircraft noise impacts, for example, is (incorrectly) associated with a prevalence rate of high annoyance of 12.3% of the population. FICON attempts no justification for its choice of 12.3% of the population as a measure of significance of noise exposure.

If, in fact, FICON (and its successor organization, FICAN) truly believed that environmental noise assessments should be based on dose–response analysis, it follows that interpretive criteria for significant noise impacts should reflect accumulating evidence from continuing progress in dose–response analyses. The 1978 "Schultz curve" was a portion of a third-order polynomial function constructed as an informal ("eyeball") fit to a collection of 161 data points derived from 18 studies. Seven of these studies (conducted with 15,430 respondents at 94 interviewing sites) dealt with opinions about aircraft noise. Quite a few additional dose–response analyses have subsequently been conducted.

3.2.4 The "Updated" Schultz Curve

FICON's 1992 "updated" curve is a univariate logistic regression function conducted on 400 of the 453 data points assembled by Fidell et al. (1989).[40] Like the Schultz curve, FICON's curve is a generic one for all sources of transportation exposure, and is not specific to aircraft noise. Including opinions about the annoyance of road and rail noise in the FICON curve has the effect of depressing the estimate of the prevalence of high annoyance due to aircraft noise, because as demonstrated by Miedema and Vos (1998), the annoyance of aircraft noise exposure is greater than that for road and rail noise at equivalent exposure levels. A simplified version of the 1998 Miedema and Vos function (Miedema and Oudshoorn 2001) was later adopted as a de facto European standard, and eventually as one of two dose–response relationships for aircraft noise in ISO 1996-1:2016.

The "updated Schultz curve" cited in the 1992 FICON report is a dose–effect function, shown in Fig. 3.3, that transforms noise exposure into a predicted annoyance prevalence rate. The FICON relationship is not a causal one, nor a theory-based one, but is merely a correlational curve fit to field measurements made prior to about 1990.

By mathematically converting noise exposure to the prevalence of a consequential degree of annoyance, this function permits algorithmic transformation of noise exposure into a tractable measure of noise effect. The function is therefore cited as providing a generic rationale for regulating aircraft noise exposure indirectly—in units of decibels—rather than as percentages of people highly annoyed.[41] Four limitations of this line of reasoning are rarely mentioned.

First, a dose–effect relationship that is anything other than a step function does not dictate any particular regulatory position. Mere knowledge of a function that

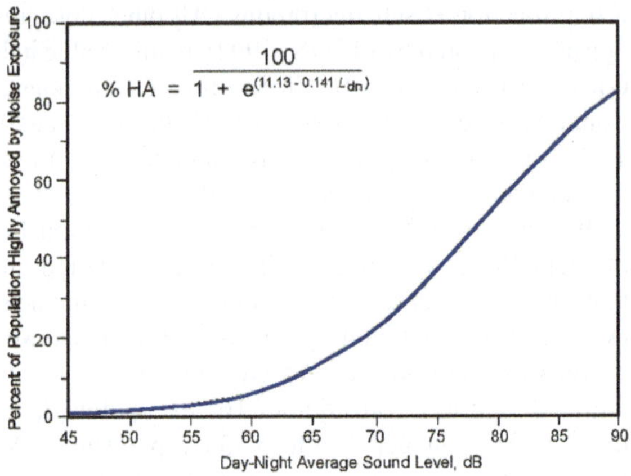

Fig. 3.3 Dose–effect function endorsed by FICON (1992) and FAA

transforms DNL values into annoyance prevalence rates does not, in itself, constitute technical, scientific, or factual justification for any value judgments about the significance of noise exposure levels that is relevant for regulatory purposes.

Second, the particular relationship (the "updated Schultz curve") that the FICON report cites is obsolete and demonstrably incorrect. Figure 3.4 compares the 1992 FICON dose–effect relationship with the latest internationally accepted relationship (ISO 2016b) for aircraft noise.

The FICON relationship under-predicts the prevalence of aircraft noise-induced annoyance of aircraft noise by a factor greater than two in the exposure region of greatest practical interest for regulatory purposes.

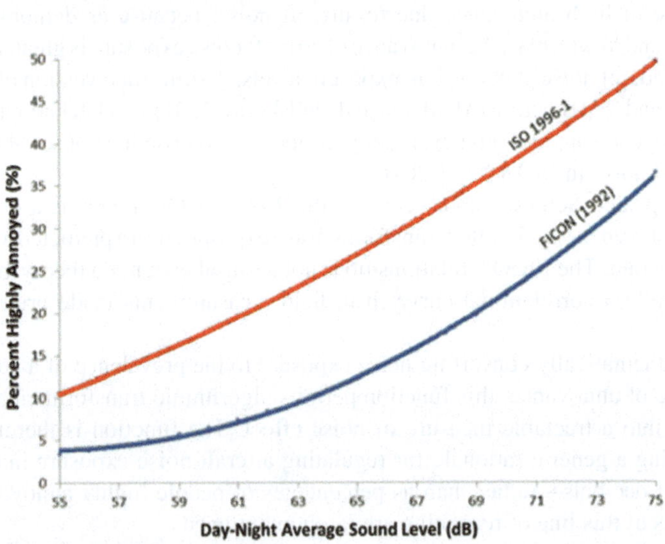

Fig. 3.4 Comparison of FICON (1992) and ISO (2016) functions for predicting the prevalence of annoyance due to aircraft noise exposure

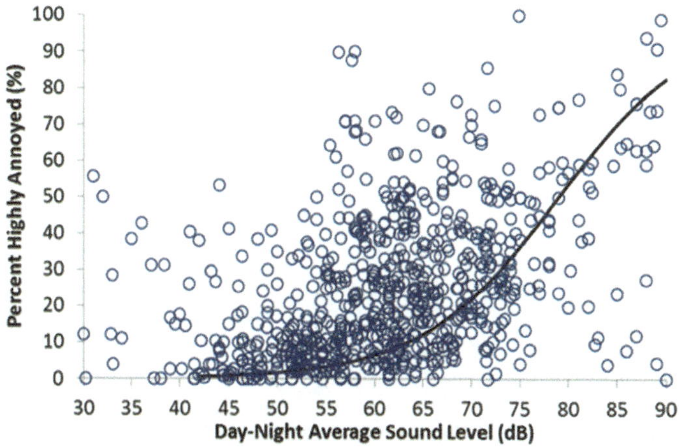

Fig. 3.5 FICON (1992) function in relation to measurements of prevalence of a consequential annoyance due to aircraft noise exposure community

Third, Fig. 3.3, which displays the FICON dose–effect function without the underlying data that it is intended to represent, misleadingly suggests that the prevalence of annoyance can be inferred from cumulative noise exposure with considerable accuracy and precision. Figure 3.5 displays the same function in relation to scores of paired field observations of annoyance prevalence rates and noise exposure. Figure 3.5 shows that the updated Schultz curve systematically under-estimates the actual prevalence of aircraft noise-induced annoyance in most communities.[42] Further, it shows that annoyance prevalence rates in different communities differ from one another so greatly that a prediction of the *average* annoyance prevalence rate for a given noise exposure level has little predictive utility when applied to any particular community.

Fourth, Fig. 3.6 shows that in the range of DNL values of greatest practical interest for regulatory purposes (55 dB $\leq L_{dn} \leq$ 65 dB), the correlation between aircraft noise exposure and the prevalence of a consequential degree of annoyance in communities is, for all intents and purposes, zero.[43] In other words, in precisely the range of DNL values in which a substantial correlation is most needed to provide strong technical support for regulatory decisions, the correlation is non-existent.

At best, a regression-derived dose–response relationship can represent the central tendency of the relationship between aircraft noise exposure and annoyance in many communities. The variability in the underlying data is so great, however, that no single, "one-size-fits-all" dose–response relationship can accurately predict the prevalence of annoyance with aircraft noise in any specific community.

Despite the above limitations, current policy is still referenced to the obsolete and incorrect FICON (1992) analysis for its positions on the significance of aircraft noise exposure. An alternative to correlation-based analyses that yields a causal dose–effect relationship and provides a more defensible rationale for policy-related purposes is described in the Sect. 3.3. The approach was originally identified by Fidell et al.

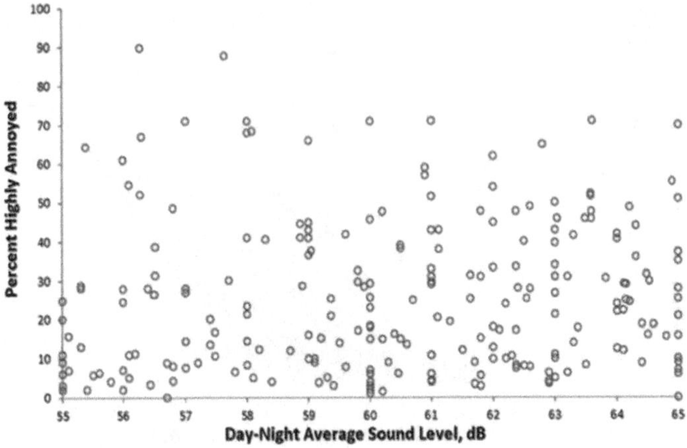

Fig. 3.6 Field observations of aircraft noise exposure and the prevalence of a consequential degree of noise-induced annoyance in communities, documenting a lack of correlation between exposure and annoyance in the range 55 dB $\leq L_{dn} \leq$ 65 dB

(1988), and subsequently elaborated by Green and Fidell (1991) and by Fidell et al. (2011).

3.2.5 *Uncertainty of Dose–Response Analyses*

A dose–response relationship is conventionally plotted on log/linear axes. Values of the independent variable (noise exposure level) plotted on the abscissa are measured in logarithmic units. The dependent variable plotted on the ordinate (the percentage of the population describing itself as highly annoyed) is measured in linear units. A 3 dB error of estimate on the exposure axis thus represents a factor of two (i.e., 100%), while a three-unit error of estimate on the annoyance axis represents just that: a 3% error. Although many consider the measurements and noise modeling predictions plotted on the abscissa to be inherently more objective and precise than the "subjective" measurements plotted on the ordinate, the reverse can be the case.

3.2.6 *Other Dose–Response Relationships*

Miedema and Vos (1998) and Miedema and Oudshoorn (2001) published separate dosage–response relationships for road, rail, and air traffic noise, derived from meta-analyses of combined data from multiple communities, not long after FICON's dosage–response relationship was published. Simplified approximations of these relationships were adopted by the European Union (2002) into a policy directive for member nations.

Fidell and Silvati (2004) and Fidell et al. (2011) suggested two other dose–response functions. The former relationship was based on nearly 53,000 interviews conducted at 326 aircraft noise interviewing sites. The latter relationship, based on nearly 76,000 interviews in 43 studies, was eventually incorporated in an international technical consensus standard (ISO 1996-1:2016). Gelderblom et al. (2017) recently compiled a database of more than 100,000 interviews from social surveys of aircraft noise annoyance conducted at 62 aircraft noise sites. The dose–response function of the latest international technical consensus standard estimates the prevalence of high annoyance with aircraft noise exposure at a value of $L_{dn} = 65$ dB as about 28%, or more than twice as great as FICON's (1992) predictive function.

3.2.7 *Other Acoustic Predictor Variables*

Mestre et al. (2011) conducted an elaborate demonstration of the high correlation of DNL with virtually all other plausible measures of aircraft noise exposure. They used INM to model aircraft noise exposure at a notional one-runway airport served by a typical fleet of aircraft. They computed values of a variety of noise metrics at a closely spaced grid of points, and then computed a correlation matrix among all of the noise metric values at these points.

The correlations between all other noise metrics and DNL were nearly perfect (typically, $r \geq 0.9$); so high, in fact, that DNL values varied from other noise metrics by little more than scale factors and constants. The very high correlations between DNL and other noise metrics mean that mathematically, univariate dose–response relationship based on noise metrics other than DNL *cannot* explain any more variance in the relationship between DNL and the prevalence of high annoyance in airport neighborhoods than DNL already does.

The only aircraft noise metrics that were not nearly perfectly correlated with DNL were threshold-type metrics, such as numbers of events in excess of a sound level, and durations of exposure in excess of sound level. Such threshold-based noise metrics share an important limitation for regulatory purposes: their slopes are so steep that regulation based on them would be essentially dichotomous. As noted by Mestre et al. (2011), the values of time-above and number-above noise metrics "…are zero until a threshold is reached, after which they climb steeply until saturation is reached…. Once the threshold is exceeded, a small change in DNL can produce large changes in [*the noise metrics time- and number-above*]. The steep slope is an artifact of the logarithmic nature of DNL but the linear nature of TA [time above] and NA [number above]". In other words, the threshold-based metrics are ill-suited for aircraft noise regulation, because they are insensitive to different degrees of aircraft noise over large portions of their ranges.

A further limitation of the threshold noise metrics for regulatory purposes is that they are sensitive to the composition of a fleet serving an airport. A fleet containing large numbers of, say, business or regional jets, but small numbers of larger jet transports, could produce the same values of a time-above noise metric as a fleet containing large numbers of much noisier four engine airliners, but very few smaller aircraft. Further, AEDT calculations of metrics such as "time-above" and "number-above" are intended for informational purposes only, and not intended as predictors of community response. FAA has no interpretive criteria for them.

DNL makes two further assumptions about the origins of annoyance with aircraft noise exposure. The first is that nighttime (10:00 PM through 7:00 AM) aircraft operations are an order of magnitude (10 dB) more annoying than aircraft operations at other times of day. The second assumption is that the optimal frequency weighting for predicting annoyance is the A-weighting network. It has been understood for several decades that the first of these assumptions is at best only approximately correct (*cf.* Fidell and Schultz 1980). In practice, however, it matters little whether an aircraft noise metric assesses an 8, 10, or a 12 dB nighttime penalty, because it is a rare civil airport that has more than about 10% nighttime operations. On a national basis, numbers of daytime operations dwarf numbers of nighttime operations, except at a small number of predominantly express delivery/air cargo airports.

Likewise, even though A-weighting of aircraft noise may not be an optimal approach to estimating its annoyance, it is certainly good enough for most regulatory purposes. A low-frequency weighting network may be preferable for predicting annoyance in special cases (such as runway sideline and other airport-adjacent neighborhoods subjected to considerable ground run-up, thrust reverser, and start-of-takeoff roll noise). A loudness level weighting[44] would probably be preferable as well, but would yield only a minor improvement in the correlation of annoyance with exposure in most cases.

3.3 Community Tolerance-Level Analysis

A minor part of the variability apparent in Fig. 3.5 may be due to various forms of error of measurement on both axes. Some of the variability may also reflect individual-level (rather than community-level) factors (*cf*. van Kamp et al. 2004). Aircraft noise in communities propagates to all individuals' residences, however, and for all practical purposes, the distributions of individual-level influences on annoyance prevalence rates in communities are unknowable in advance. Individual-level explanatory factors (covariates in multi-level regression analysis) are thus of meager utility for regulatory purposes.

The recently standardized (ISO 1996-1:2016) Community Tolerance Level (CTL) analysis represents a paradigmatic shift (per Kuhn, 1962) in the analysis of empirically observed differences in annoyance prevalence rates in communities exposed to similar levels of transportation noise. CTL distinguishes acoustic from non-acoustic influences on annoyance prevalence rates, and attributes much of the variability in annoyance prevalence rates among communities to community-level factors. The

CTL approach to analyzing social survey data from different communities is summarized in a simple set of equations that may be found in Appendix A of Fidell et al. (2014). The basic predictive relationship is:

$$p\,(HA) = e^{(-A/m)} \tag{1}$$

where:

$p\,(HA)$ is the proportion of the population highly annoyed by noise exposure,
A is a community's annoyance decision criterion[45], and
m is the noise dose, defined as:

$$m = \left(10^{(L_{\text{dn}}/10)}\right)^{0.3} \tag{2}$$

and where:

L_{dn} is the Day–Night Average Sound Level, in decibels.

Empirically, the proportion of a community highly annoyed is determined in field settings by direct questioning in a structured social survey interview.[46] The functional relationship specified by the basic predictive relationship of Eq. (1) is a sigmoid. The exponential form of the relationship was chosen as the most plausible and parsimonious (i.e., single parameter) transition function to model the growth, from zero to one, of the proportion of a community highly annoyed by increasing levels of noise exposure.

CTL analysis treats the proportion of a community that describes itself as highly annoyed as equally influenced (1) by noise exposure per se, and (2) by a non-acoustic criterion for self-reporting of annoyance. Univariate regression analyses (those based on a single predictor variable, noise exposure) ignore the influence of the second of these determinants of annoyance prevalence rates. In CTL analysis, the slope of the dose–response relationship is fixed at that of the exponential growth rate of loudness with sound level, while the parameter A in Eq. (1) translates the sigmoidal function back and forth along the abscissa.

A family of CTL curves may be seen in Fig. 3.7 over the range of DNL values that are of primary regulatory concern. All of the curves are of the form $e^{(-x)}$. They are not parallel within the plotted range of DNL values because shifting them left and right along the abscissa displays different portions of the sigmoids in the plotting window.

In practice, the value of A for a particular community is determined empirically, from social survey findings reported as the proportion of a community's residential population describing itself as highly annoyed by noise exposure for a given DNL value. A Community Tolerance Level is defined by convention[47] as the value of DNL at which 50% of the population is expected to describe itself as highly annoyed by noise exposure. CTL values may be calculated from the value of A that produces the smallest RMS error,[48] as shown in Eq. 3:

$$L_{ct} = 33.3 log_{10}(A) + 5.32 \tag{3}$$

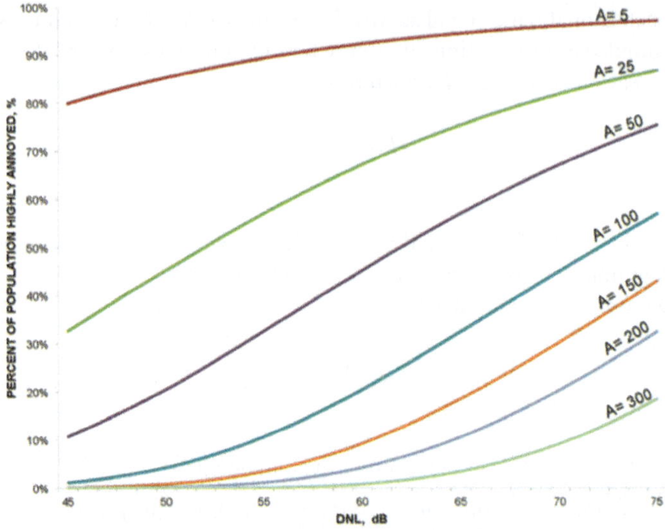

Fig. 3.7 A family of dose–response functions, parametric in A (the non-acoustic response criterion), relating cumulative 24-h noise exposure to the prevalence of a consequential degree of transportation noise-induced annoyance in a community

The panels of Fig. 3.7 show sample fits of social survey data to the CTL function for six airports. These fits are *not* derived by any correlational curve fitting method, but are simply those achieved by shifting an effective loudness function along the abscissa. Additional examples may be found in Fidell et al. (2011) and Schomer et al. (2012) (Fig. 3.8).

If CTL for a community is known, then the annoyance decision criterion, *A*, may be calculated from CTL as shown in Eq. 4:

$$A = 10^{\left(\frac{L_{ct}-5.32}{33.3}\right)} \quad (4)$$

The distribution of *A* for annoyance due to aircraft noise over communities is approximately log normal, as shown in Fig. 3.9, adapted from Fidell et al. 2011. (Alternatively, the distribution is Gaussian in log(A).) Most communities show relatively little tolerance for exposure to aircraft noise, but a few are highly tolerant of it.

For national regulatory purposes, however, the exact combination of reasons that particular communities are intolerant of aircraft noise is a matter of idle curiosity, or of secondary concern at best. Aircraft overfly residents of all types in neighborhoods near airports—men and women, young and old, sensitive and insensitive—regardless of their attitudes toward airport operation. Adopting a nationwide regulatory strategy that seeks to protect the *individuals* residing in all communities who are most likely to be highly annoyed by aircraft noise would lead to unnecessarily stringent policy thresholds in most communities. The obvious compromise and regulatory "sweet

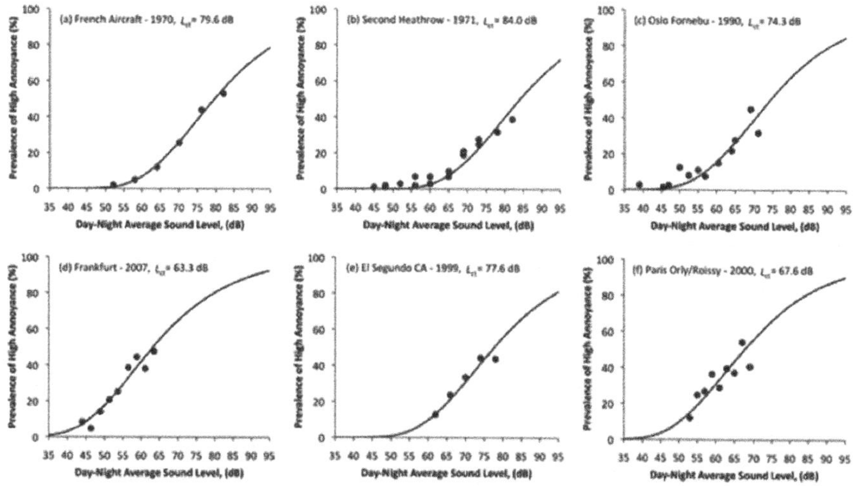

Fig. 3.8 Examples of fit of field measurements of prevalence of high annoyance to effective loudness curve for six studies of community reaction to aircraft noise

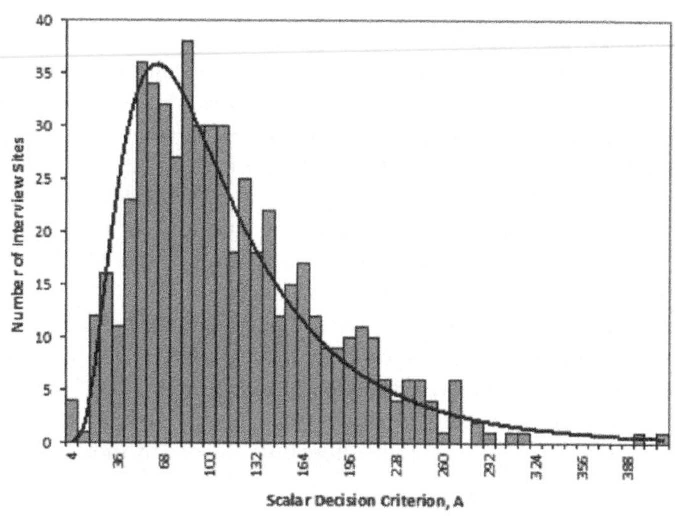

Fig. 3.9 Distribution over communities of the scalar quantity A in CTL analysis, for annoyance due to aircraft noise

spot" is a national policy strategy flexible enough to yield a constant prevalence rate of high annoyance among residents of individual *communities,* or perhaps within classes of communities.

3.4 Time Constants of Arousal and Decay of Annoyance

Little is known quantitatively about how long it takes for annoyance prevalence rates to reach asymptotic values following exposure to increased noise exposure, nor about how much time must pass following a decrease in noise exposure before annoyance subsides to baseline rates. On the basis of repeated interviews undertaken in conjunction with runway repairs at one airport, Fidell et al. (1985) estimated that the time constants of arousal and decay of annoyance are on the order of at least weeks, if not months. Horonjeff and Robert (1997) consider the evidence to be inconclusive for a variety of reasons, including a large amount of scatter in the results of a small number of field studies, a lack of sophistication in analyses, and the limited number of opportunities to observe changes in annoyance prevalence rates accompanying major changes in aircraft noise exposure levels.

The issue of time constants of community annoyance is not merely of academic interest, because aircraft noise exposure in some cases (e.g., at airports with multiple, non-parallel runways) can vary seasonally, as do weather, repair work, and consequent runway use patterns. At large civil airports, such swings in noise exposure levels were more common in former years than at present, when differences between weekday and weekend flight schedules are generally minor.[49]

FAA mandates characterization of civil aircraft noise on an "annual average day" basis for environmental impact analyses of aircraft noise, rather than over any shorter term, or more specific time period. The annual average day is *not* any actual day, but a hypothetical day when all of the factors that affect aircraft noise exposure in airport environs—runway use, fleet mix, flight stage lengths, flight paths, wind speed and direction, temperature and humidity, and so on—simultaneously assume annual average values. Aircraft noise exposure contours produced for an "annual average day" of noise exposure at an airport may therefore never be actually experienced by residents of airport neighborhoods on any actual day.

Reliance on annual average day estimates of neighborhood noise exposure tacitly assumes that the prevalence of annoyance in a community is not very volatile. On the contrary, annual average day estimation of exposure to aircraft noise assumes that community response to aircraft noise responds only sluggishly to day-to-day variation in airport operations, and is therefore little affected by short-term variability in airport operations.

3.5 Complaints

Aircraft noise complaints are behaviors, such as telephone calls made to airport noise offices or complaint reporting hot lines, or visits to web-based complaint collection websites. (More recently, automated means have been developed for registering complaints digitally.) Complaint behavior differs from annoyance in that complaints are volunteered by self-selected complainants, rather than systematically solicited from a noise-exposed population. As such, complaints are not necessarily representative of any specifiable population. Since the prevalence of a consequential degree of noise-induced annoyance is FAA's preferred measure of aircraft noise impact, complaints are viewed at a federal level as unreliable and uninterpretable for most purposes.[50]

This view has led to a disconnect between federal policy and local practice in gauging community reaction to airport noise. For federal purposes related to NEPA-mandated disclosures of aircraft noise impacts, complaints play no role whatsoever. For purposes of ongoing local assessments of community reaction to airport noise, many airport managers are content to focus on "How many noise complaints has the airport received for each of the last few months?" Despite relying on complaint tallies for at least a rough indication of airport-community relations, most airports denigrate noise complaints for another reason: small numbers of complainants usually account for large percentages of all complaints.[51]

Airport noise offices processed noise complaints rather simplistically for many years. Complaint analyses often amounted to little more than sticking pins in maps at the addresses of complainants, periodically tallying and reporting complaint numbers by neighborhoods, and speculating about geographic relationships between aircraft noise exposure lobes and concentrations of complainants. As more capable airport noise and operations monitoring software developed over the last decade or so, efforts to interpret complaint data have become more sophisticated.

For example, Fig. 3.10 depicts the spatial complaint density (complaints per unit area) behind San Francisco International Airport's two main departure runways (the 01 pair). The peaks of the complaint density correspond with the 45° lobes (135° to both sides of an aircraft's nose) of engine noise behind departing aircraft. The offset of the complaint density peaks may reflect the prevailing westerly wind.

For another example, Figs. 3.11 and 3.12 show spatial complaint densities in year-long periods prior to and after the opening of a newly constructed parallel runway at Seattle-Tacoma International Airport. The third runway was built a few hundred yards from two existing runways, and immediately put into service to permit repairs

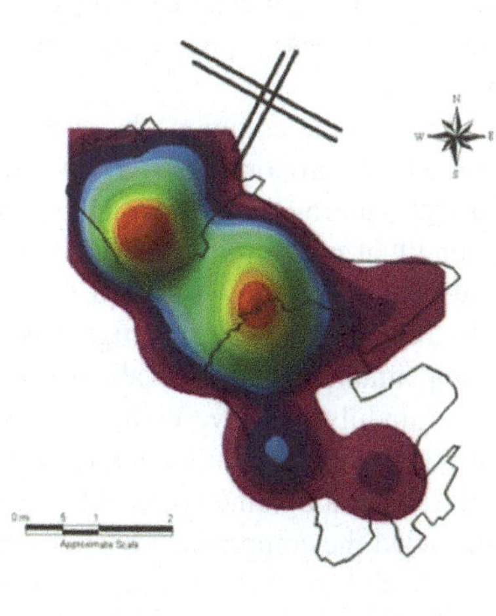

Fig. 3.10 Correspondence between high spatial complaint densities and directivity pattern of engine noise from aircraft taking off from SFO's main departure runways

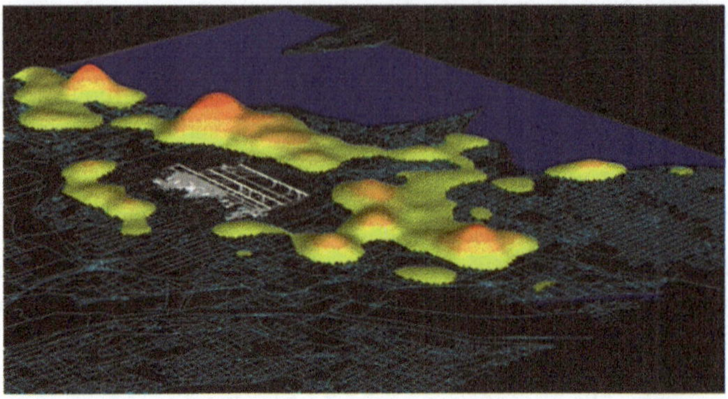

Fig. 3.11 Three dimensional spatial density representation of complaints in twelve months prior to the start of operations on a new runway

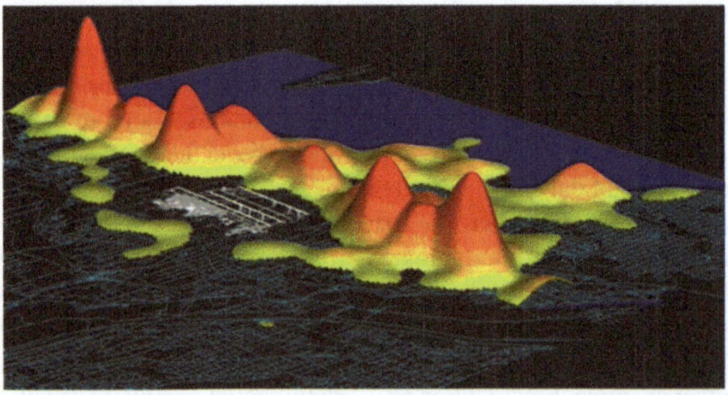

Fig. 3.12 Three-dimensional spatial density representation of complaints in twelve months following the start of operations on a new runway

to be made to an existing runway. Due to the proximity of the new runway to the older ones, noise exposure levels in the community to the west of the airport changed very little.

Despite the common observation that small numbers of complainants account for a large proportion of complaints, complaint rates have been found to follow lawful patterns. As shown in Fig. 3.13, Fidell, Mestre and Sneddon (2012) have found that numbers of complaints lodged per complainant closely follow a power law relationship known as Zipf's Law. (Very large numbers of annual complaints may also be expected due to the highly skewed, "tail-heavy" distribution of complaints per complainant.) Complaints about aircraft noise are sometimes dismissed as originating from unduly sensitive complainants. "You shouldn't have chosen to live near an airport if you couldn't tolerate aircraft noise" is among the more benign of such retorts. Although the comment assumes that the noise created by airport operations preceded the complainant's tenure of residence, it is changes in airport runway use,

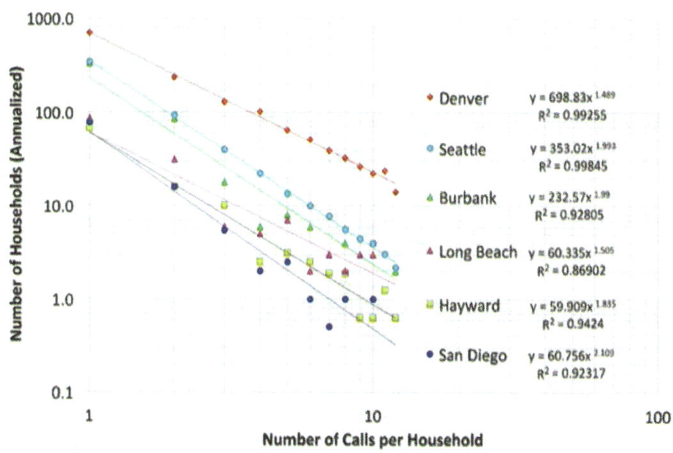

Fig. 3.13 Power law behavior of complaints per complainant at half a dozen airports

increases (especially in nighttime air cargo) operations, or more recently, changes in approach and departure procedures, that commonly lead to complaints.

Figure 3.11 shows that relatively small numbers of complaints per unit area were lodged from homes immediately to the west of SEA's north/south runways before the third parallel runway was built. In contrast, in the year after the new runway opened, Fig. 3.12 shows that the spatial density of complaints received from homeowners living near it on the west side of the airport increased greatly.[52]

3.6 Speech Interference

Residential speech interference (difficulty understanding face-to-face and telephone conversation, and masking of radio and television speech) is among the more frequent complaints about aircraft noise in airport neighborhoods. Episodic interference with indoor and outdoor speech by urban noise in general, and by the noise of arriving and departing aircraft in particular, is perhaps the most common complaint about transportation noise. In airport neighborhoods near runway ends, each flight operation may interfere sufficiently with speech to introduce forced lapses in conversation hundreds of times a day, each one lasting as long as five to ten seconds. Such interference can be exacerbated in areas underneath precision-based navigation routes, which direct one aircraft after another over the same homes for hours at a time, at intervals as short as every two minutes.

When the task at hand is verbal communication, frequent interruptions of speech may be viewed as a form of task interference. In face-to-face conversation, such interruptions are usually accommodated by some combination of three coping strategies: (1) raising the level of vocal effort, (2) reducing the communication distance, or

(3) pausing until the flyover noise abates. Not all of these strategies are practical or effective in non-residential settings, such as school, church, or other public address circumstances. Delays and forced lapses in verbal communications caused by flyover noise can be highly distracting, and hence annoying. The simple correlation between the incidence of reported speech interference and of high annoyance is in excess of 0.80 (Fidell 1978).

Figure 3.14 summarizes the ability of steady, broadband noise to reduce sentence intelligibility. At A-weighted single-event sound pressure levels below about 60 dB, noise has little effect on conversations conducted with relaxed vocal effort (i.e., at indoor speech levels on the order of 65 dB within three feet of a speaker). Sentence intelligibility does not decline noticeably until the level of masking noise approaches or exceeds 70 dB, and even then, degrades only gradually until masking noise levels reach the high-60 dB range. Context alone provides sufficient redundancy to maintain sentence intelligibility even in the presence of moderate levels of background noise. Isolated word intelligibility degrades more rapidly than sentence intelligibility with increasing masking noise level, but is not the main concern of most airport-vicinity residents.

Despite the prominence of speech interference as a source of annoyance with aircraft overflights, FAA has no specific criteria for gauging the significance of aircraft noise-induced speech interference. For example, FAA offers no interpretive criteria couched in terms of tolerable numbers of episodes of interrupted speech, or total duration of speech interference during an annual average day of aircraft noise exposure.

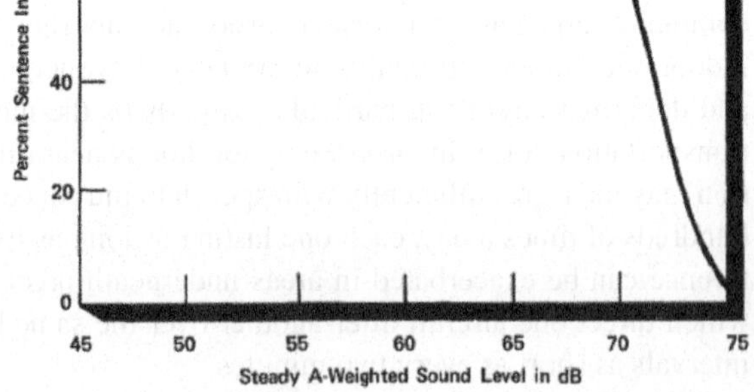

Fig. 3.14 Relationship between sentence intelligibility and steady-state ambient sound levels

3.7 Sleep Disturbance

Sleep disturbance caused by aircraft noise is a familiar phenomenon to many, but a difficult one to rigorously define, accurately predict, and meaningfully interpret. This difficulty reflects in part the lack of full agreement among researchers on measures and definitions of sleep disturbance. Such disagreement extends to matters as basic as the size and character of effects observed in laboratory and field studies; the role of long-term habituation to nighttime noise in familiar sleeping quarters; the health consequences of sleep disturbance; and the most appropriate measures and methods for characterizing and analyzing sleep disturbance.

Fidell et al. (1995) note that the credibility, generalizability, and utility of sleep disturbance predictions are also limited by small and non-representative samples of test participants, and by restricted (airport-specific and relatively short duration) circumstances of exposure. Although expedient relationships may be the best available, their predictions are of only meager utility for policy analysis and regulatory purposes: they account for very little variance in the association between environmental noise and sleep disturbance, have characteristically shallow slopes, have not been well validated in field settings, are highly context-dependent, and do not squarely address the roles and relative importance of non-acoustic factors in sleep disturbance. Such relationships offer the appearance more than the substance of precision and objectivity.

Laboratory studies of sleep tend to rely on electrophysiological measures of sleep disturbance, while field studies often rely more on behavioral indications of sleep disturbance, such as motility and behaviorally confirmed awakening (Basner and Samel 2006; Michaud et al. 2007).[53] Other difficulties in defining and predicting noise-induced sleep disturbance include the expense and complexity of large-scale, long-term field studies, non-representative self-selection of test participants, and the lack of a large database of field observations of sleep disturbance.

No U.S. standard currently predicts noise-induced sleep disturbance from outdoor measurements of aircraft noise exposure. A 2008 American National Standards Institute standard (ANSI/ASA S12.9-2008/Part 6) was recently retracted for reasons discussed at length by Fidell et al. (2018). Briefly, these reasons included questionable statistical assumptions made in a meta-analysis of a relatively small corpus of field studies, the lack of generalizability of findings from one airport to another, and the poor strength of association of the relationship developed for linking aircraft noise exposure to sleep disturbance. Furthermore, there is good reason (Fidell et al. 2013) to believe that sleep disturbance in residential settings is more closely linked to familiarity with the noise environment in sleeping quarters than with absolute levels of aircraft noise intrusions.

3.8 Task Interference

Aircraft noise in airport-vicinity residences can obviously interfere with tasks that depend on unmasked auditory feedback to sustain performance. Such tasks are not limited solely to those that require continuity of speech intelligibility (e.g., conversation, telephone use, listening to radio or television, as discussed in Sect. 3.6), but can also include those which require timely, non-speech information to sustain performance. For example, transient auditory cues such as beeps, clicks, and other sounds produced by user interfaces to computer software can be subtle and easily masked by aircraft overflights. Even if largely unacknowledged in highly practiced tasks, such routine auditory feedback can be important in acquiring skills.

Although noise can interfere with performance of non-auditory tasks (Kryter 1984), the effects of aircraft noise on task performance in airport-vicinity residences are too diffuse to support inference of a dose–effect relationship. Familiar aircraft flyover noise is unlikely to interfere directly with non-auditory task performance in residential settings.[54] Apart from masking auditory cues that may provide task performance feedback, aircraft noise at levels typical of indoor exposure are unlikely to interfere with many routine, highly-practiced, low-skill, everyday residential tasks. Aircraft noise in airport neighborhoods is hardly unfamiliar or unpredictable, and thus is unlikely to occasion much startle.

3.9 Extra-Auditory Health Effects

Epidemiologic and other research on potential effects of aircraft noise on health has been conducted since the early 1950s (Benox 1953). The initial technical consensus—that the only direct risk (outside of occupational settings) posed by aircraft noise exposure is of hearing damage—was largely unchallenged for two decades (Harris 1994). This consensus began to erode in the 1970s as commercial jet transportation became more widespread; as societal concern for environmental pollution became widespread and analytic research tools became more convenient; and as U.S. and later European funding became available for epidemiologic studies. Half a century later, the original consensus is widely regarded as ill-informed, even though the issue of extra-auditory health effects of noise exposure remains controversial, and far from resolved.

3.9.1 Relative Risks of Noise-Induced Health Effects

If research findings demonstrated substantial relative risks[55] of putative extra-auditory health effects, much of the controversy would be dispelled. Such is not the case however. A recent World Health Organization publication (WHO, 2018) publication rated information about the effects of aircraft noise on cardiovascular health relative risks as "very low quality" (for ischemic heart disease) and "low quality" (for hypertension). WHO nonetheless inexplicably "strongly recommends reducing noise levels produced by aircraft below 45 dB L_{den}, as aircraft noise above this level is associated with adverse health effects". (As noted below, WHO's position is quite controversial.)

Relative risks analyzed by WHO in studies of both ischemic heart disease and hypertension had very wide confidence intervals, which in some cases were lower than 1.0 (implying that aircraft noise exposure *protects* against risks of cardiovascular disease). The very longevity of attempts to demonstrate meaningful extra-auditory effects of aircraft noise exposure is an indication of the absence of persuasive evidence of a meaningful effect (or alternatively, of the subtlety and complexity of the issue).

3.9.2 Health-Related Justification for Aircraft Noise Regulation

As a self-evident generality, most people don't like aircraft noise. Were it not for the equally self-evident economic utility of air transportation and consequent demand for air transportation services, such dislike alone would be adequate grounds for regulation of aircraft noise. After all, society routinely regulates disfavored behaviors and conditions, small and large, and of all kinds, ranging from graffiti, litter, strong odors, public nudity, commercial signage, panhandling, disturbing the peace, unfair business practices, racial and sexual discrimination, and so on, *without* any pretext that such conditions and behaviors pose any meaningful hazards to public health.

Why, then, do many believe that evidence of public health hazards is essential to the rationale for aircraft noise regulation? Most routinely-regulated, non-health-related actions offer little or no tangible benefit to society as a whole, cost little to implement, and do not infringe greatly on commercial interests. The costs of aircraft noise regulation, on the other hand, are seen as great enough, and to affect enough people, that they outweigh the costs of merely annoying a relatively small number of people. Stated more directly, the costs of risking the health of even relatively few people are apparently great enough to some to justify imposing economic penalties on others.[56] This judgment is an explicitly non-technical one, about which different societies may disagree. U.S. and northern European countries differ notably on the importance that they attach to potential adverse health effects.

Some proponents of aircraft noise regulation seem compelled to press public health-related arguments as a rationale for stricter regulation than seems necessary to aviation industry interests. In the United States, the contention may be traced at least as far as charters and policy preferences of EPA's Office of Noise Abatement and Control ("EPA/ONAC") and FAA's Office of Environment and Energy (FAA/AEE). EPA/ONAC, established under the Noise Control Act of 1972, was authorized to identify noise sources and regulate them in the interests of public health and welfare.

During the same time period, FAA's charter from Congress included promotion of civil aviation. The conflict in the charters of the two federal agencies reached a peak during the late 1970s, at which time EPA/ONAC actively supported studies of potential extra-auditory health effects of noise exposure in animals and humans.[57]

The conflict between promoting industry interests and regulating in the interests of public health and welfare was resolved in favor of industry in the United States when EPA/ONAC's 1981 budget was reduced to zero. By then, however, Northern European enthusiasm for health-related noise effects research had begun to increase, eventually yielding a stream of large-scale epidemiological studies (among them, Knipschild 1980; Knipschild et al. 1981; Neuss et al. 1983; Babisch and Gallacher 1990; Passchier-Vermeer et al. 2007; Babisch et al. 2008; zur Nieden et al. 2016). These and other studies are commonly cited as providing evidence of a range of alleged adverse physical and mental health consequences of environmental noise exposure.

3.9.3 Hypothesized Mechanism of Disease Induction

No direct neural pathway capable of directly or promptly inducing cardiovascular disease links the cochlea to the heart. Instead, putative adverse effects of high-level noise exposure on cardiovascular health are hypothesized to be slowly mediated, over the course of years, by sustained high levels of stress hormones such as adrenaline and cortisol.[58] Noise exposure is hardly the only source of stress in modern urban life, however (*cf.* Glass and Singer 1972). Modern civilization is replete with everyday stressors, such as crowding; air pollution; occupational, social, and financial challenges; and diffuse anxieties of all sorts, to mention just a few.

No evidence-based quantitative estimates are available of the relative contributions of aircraft noise exposure and life's other stressors to total individual stress loads. It is therefore impossible to definitively determine what proportion of stress-related disease can be attributed to aircraft noise per se. Exposure to high levels of aircraft noise may be a major source of stress in some people's lives, but may be dwarfed by other sources of stress in other people's lives. Given such uncertainty, it is difficult to compose an objective, health-based rationale for aircraft noise regulation.

3.9.4 WHO's Recent Recommendations

The most recent summary of extra-auditory health findings is that of the European Regional office of the World Health Organization (WHO 2018). The technical analyses supporting these recommendations have been challenged by Gjestland (2018). WHO's recommendations proceed from absolutist definitions of public health, and of the responsibilities of governments for protection of citizens from health risks. The report adopts the definition of health of WHO's Constitution: "Health is a state of complete physical, mental and social well-being and not merely the absence of disease or infirmity" (WHO 1946). It also adopts the European Union's belief that "no person should be exposed to noise levels which endanger health and quality of life".

Neither of these perspectives is particularly well suited to the U.S. regulatory environment. WHO's definition of "health" is a highly bureaucratic one devised in the 1940s to maximize the scope of a nascent U.N. agency's remit to address any condition or circumstance that the agency interprets as health-related. The notion that national governments' proper role is to assure that no citizen should ever be exposed to noise levels which endanger complete individual well-being or quality of life is a similarly uncompromising one. Noise regulation in the United States is intended to balance conflicting societal interests. Such resolution does not involve substitution of regulatory fiat for individual judgment (for instance, about choice of place of residence). Instead, regulation in the United States inevitably entails compromises among societal costs and benefits of regulation, and often between the interests of producers and consumers of noise pollution.

WHO asserts that its recommendations are not based solely upon the quality of technical evidence, but also upon a non-technical set of value judgments. These include, among others:

Resource implications—the apparent cost-effectiveness and benefits of regulatory action;
Environmental justice (equity and human rights considerations)—the greater the likelihood that regulatory action will reduce inequities or contribute to the realization of human rights, the greater the likelihood of a strong recommendation;
Acceptability—the greater the acceptability of a (regulatory) option to stakeholders, the greater the likelihood of a strong recommendation;
Feasibility—the more feasible an option seems to stakeholders, the greater the likelihood of a strong recommendation.

This is an odd set of value judgments on which to base technical recommendations, since it includes several elements of appearances and public opinion. It is nonetheless important to understand the nature and limitations of the technical aspects of WHO's recommendations.

3.9.5 Nature of Epidemiologic Evidence

Hill (1965) identified a set of criteria that have since become the conventional wisdom for imputing causality from epidemiologic evidence. These criteria included the strength and specificity of any association between a purported cause and its effect; temporal appropriateness (causes must precede effects); a credible dose–response relationship; a plausible mechanism of causal action; consistency between laboratory and field findings; and a demonstration of the effectiveness of empirical intervention.

Weak associations between a potential cause and its putative effect, often excused by a range of uncertainties of measurement and confounding factors, are not credible evidence of causality. Geographic associations are the weakest of all forms of epidemiologic evidence, such as one between area-wide estimates of aircraft noise exposure and the incidence of some disease or condition. The bulk of field studies on extra-auditory effects of aircraft noise are of precisely this sort. They include, for example, cross-sectional comparisons of mortality and morbidity incidence rates in areas around airports, contrasted with those observed elsewhere.

The proper level of epidemiologic analysis is the individual, as in longitudinal and case–control study designs. Without information about personal noise exposure and individual durations of residence, simple comparisons of records of disease incidence, hospital admissions, death certificates and the like in areas near airports and supposed control areas with putative lesser levels of aircraft noise exposure reveal little or nothing about any causal relationships between aircraft noise and disease.

For reasons discussed by Thompson et al. (1989), various forms of cardiovascular disease are among the more commonly investigated potential extra-auditory health effects of aircraft noise exposure. Cardiovascular disease does not develop overnight however, but rather with a latency (induction) period of at least a decade. Simple geographic associations between cardiovascular disease and place of residence that do not take into consideration personal (rather than place) noise exposure and duration of residence reveal little about any potential causal relationship between aircraft noise and cardiovascular health. Airport infrastructure and flight operations only rarely remain constant over periods of many years, nor do large proportions of airport-area residents remain in the same homes for similar time periods. Gross geographic associations thus do not meet Hill's (1965) criterion for "appropriateness".

Morbidity and mortality studies are even less useful as indications of adverse health effects of aircraft noise exposure, for simple reasons of logic. Suppose, for instance, that it were discovered that greater-than-expected numbers of deaths attributed to cardiovascular disease occurred in an airport neighborhood than in a neighborhood with lower aircraft noise exposure. Since everyone eventually dies of some adverse health condition, greater-than-expected numbers of deaths attributable to cardiovascular disease imply lesser-than-expected numbers of deaths attributable to other diseases—say, cancer. It makes no more sense in this case to infer that aircraft noise "causes" cardiovascular disease than to infer that aircraft noise "protects" against cancer.

Consider, for example, the Jones and Tauscher (1978) claim of a geographic association between aircraft noise exposure and a birth defect (polydactylism). Jones and Tauscher found a higher incidence of reportable birth defects in census tracts partly or wholly within an idiosyncratically defined peak single-event aircraft noise contour at Los Angeles International Airport than in the remainder of Los Angeles county. The effect itself—six excess cases—was miniscule, but arguably (barely) distinguishable from an effect that could have arisen by chance alone.

Bader (1978) points out that birth certificate data are subject to such wide variation due to reporting practices that the Jones and Tauscher findings are not meaningful in the first place. Using similar birth certificate data, Bader demonstrated that the rates of birth defects in cities near the Seattle-Tacoma airport were even lower than rates elsewhere in the region. Edmonds et al. (1979) conducted a similar study for the Atlanta airport using data from the Metropolitan Atlanta Congenital Defects Program, which employs multiple methods of case ascertainment. They found no differences in the rates of 17 categories of defects between high ($L_{dn} \geq 65$ dB) and low noise ($L_{dn} < 65$ dB) census tracts after controlling for hospital of birth, socio-economic status and race. A matched case–control study of all neural tube defects likewise showed no statistically significant association between supposedly high noise exposure and neural tube anomalies.

Studies such as that of Jones and Taucher (1978) that produce geographic associations between aircraft noise exposure and adverse health consequences typically fail Hill's criteria for causality on grounds of replicability, biological plausibility, specificity, and appropriateness of latent and induction periods. Polydactylism, one of many teratogenic conditions, is a genetically determined condition which lacks any plausible biological linkage to aircraft noise exposure. The findings of Jones and Taucher (1978), like those of many similar studies of simple geographic associations between noise and health (e.g., Meecham and Smith 1977, Meecham and Shaw 1979), are thus of essentially no relevance or utility for aircraft noise regulatory purposes.

Logical obstacles such as those noted above to inference of causal relationships between aircraft noise and adverse health consequences are routinely ignored in the popular press, and even in some scholarly discourse. Review articles commonly include statements that may be paraphrased as "No single study has revealed evidence of strong causal relationships between aircraft noise exposure and extra-auditory health effects, although many have found weak but suggestive evidence of adverse effects. Because public health is such a consequential matter, it is best to err on the side of caution when formulating noise regulatory policy". This is hardly a balanced, detailed, or carefully reasoned rationale for regulation. It is, rather, a rationale about which national regulatory bodies may reasonably disagree. Not even a tall pile of studies yielding weak or otherwise "suggestive" evidence of adverse health effects of residential aircraft noise exposure provides a compelling rationale for regulation.

3.9.6 Limitations of Dose Measurement

Perhaps the most obvious inadequacy of common metrics of environmental noise exposure for epidemiologic research purposes is that they are inherently place-oriented metrics that may bear little actual relationship to personal exposure. Worse yet, outdoor exposure levels are the only ones that can be measured or predicted at reasonable expense and defended as even arguably representative of the noise exposure of residential populations. Granting for the sake of argument that the outdoor noise exposure of a residential neighborhood can be expressed with useful precision in numeric form, how much can one predict about the noise exposure produced inside particular residences? How much can one predict from outdoor neighborhood noise measurements about personal noise doses of individual household residents?

At first glance, it might appear that indoor and outdoor noise exposure would differ only by a constant (the transmission loss of a typical residential structure) and a small error term. If true, this difference would introduce some uncertainty into predictions of indoor exposure, and the resulting uncertainty could be viewed as just another source of bias or error variance. Unfortunately, matters are not this simple. Even if one could estimate indoor exposure levels with useful accuracy and precision, one could not lay claim to a persuasive metric of personal exposure for two additional reasons:

(1) The indoor noise environment of a residence bears little direct relationship to the outdoor noise environment of a neighborhood, because the indoor noise environment contains its own noise sources: household equipment and appliances, radio and TV, and other sounds of human habitation. Thus, the level of the indoor noise environment of residences may be higher or lower than the level of the outdoor noise environment at different time periods throughout the day (Fidell et al. 2013); and
(2) Household residents are not stationary objects; they move about within homes and leave home entirely for long periods during the day and night. At such times, no correlation whatever exists between personal noise exposure and outdoor neighborhood noise.

3.10 Geographic Distribution of Costs and Benefits of Air Transportation

Aircraft noise is what economists consider a "negative externality"—a cost that is not fully priced into passenger airfares, air freight charges, and airport operating budgets. The costs to communities of airport proximity can be substantial, and greatly out of proportion to any local benefits. They include pressures on—if not actual reductions in—municipal real estate tax bases, property values, and sales tax revenues; loss of control over the character of community development; diminution of quality of life for noise-exposed residents; additional hindrances to local jurisdictions such as constraints on the quality and growth of municipal services; and competition for issuance of municipal bonds.[59]

3.10.1 Uneven Distribution of Costs

All other things being equal, the hinterlands of large metropolitan airports extend halfway to the nearest other major airport.[60] For example, Californians living between Los Angeles and San Francisco (338 air miles apart) who wish to fly to overseas destinations have few convenient air travel choices other than an airport in one of the two metropolitan areas. Very few of the hundreds of communities and millions of travelers who enjoy the benefits of operation of airports in these two metropolitan areas are subjected to high levels of residential noise exposure created by aircraft arriving at and departing from them. On the other hand, tens of thousands of people living in a small number of communities close to airports bear the brunt of the aircraft noise exposure that benefits travelers throughout the two airports' hinterlands.

The manner in which the aircraft noise regulatory environment has developed has created a disproportionate distribution of the noise-related costs of airport operation throughout airports' hinterlands. A high noise-level threshold for the definition of significant aircraft noise exposure, for example, minimizes the necessity for disclosing the adverse environmental impacts of airport operation in nearby communities, and thus facilitates construction of airport infrastructure such as runways, terminals, and ramp space for engine maintenance and night air cargo business.

Airline economics are responsible for some of this disparity in costs and benefits. The operating efficiencies of concentrating air travel services in a small number of airports in a given region provide airlines with greater market access through greater numbers of flight connections. Greater flight frequencies, in turn, facilitate greater aircraft utilization, higher passenger revenue, and spread the costs of gate, terminal, and other airfield services over a wider base. Greater numbers of enplanements also imply greater sales and hotel tax, as well as other special facility fees paid to local governments. Airport, airline, and sometimes even local government interests are thus aligned with respect to the economic efficiencies of scale.

Such efficiencies of air transportation, however, contribute to a diminished quality of life, smaller property tax bases, constraints on non-transportation-related community services, and limited opportunities for residential real estate development in airport-vicinity communities. In some cases, communities may decide that the imbalance in the costs and payoffs of airport and non-airport-related land uses implies that airports do not invariably constitute the best and highest use of valuable urban land.

3.10.2 Airport Economic Impact Assessments

According to the U.S. Department of Transportation (2016), the nation's 3300+ civil airports support 11 million aviation industry jobs, and are responsible for $1.6 trillion in total economic activity. Individual airport-sponsored economic impact analyses must nonetheless be taken with a grain of salt. Airport managements tend to conflate their own interests with those of all surrounding communities, even when airport operations create more noise in some communities than in others. Airport economic impact analyses typically are unsophisticated, exaggerate the economic contributions of airports to regions, are designed and paid for by aviation project proponents, and invariably claim credit for regional economic success. Boon and Wit (2005) catalog typical flaws of airport economic impact analyses. Such analyses focus on total economic benefits of airport operation, rather than marginal economic impacts of specific airport improvement projects. They mis-estimate numbers of jobs created by aviation projects by assuming that aviation workers would be unemployed if such projects were not undertaken; and they often sum direct and indirect jobs across job sectors, leading to over-counting and over-estimation of total employment.

They rarely distinguish between cause and effect, often ignore aviation-related costs, and rarely conduct credible, formal cost/benefit analysis. When all is said and done, it remains unclear whether communities thrive because they happen to be located near busy airports, or whether airports thrive because they happen to be located in economically successful communities.

The usual analytic approach estimates direct employment and spending, and then applies multipliers to account for indirect and induced economic activity. The approach is susceptible to double-counting, dubious claims of induced economic activity (as, for example, taking credit for aviation-related industrial manufacturing that happens to be located within an airport's hinterland), and takes credit for transfers from one geographic region to another. ACRP Report 132 (Economic Development Research Group et al. 2015) describes alternate approaches to assessing airport economic impacts.

Although the costs of noise exposure may arguably balance the benefits of air transportation over large geographic areas, communities near airports pay the price in aircraft noise exposure for benefits enjoyed by residents of communities farther from airports. In contrast to the early days of aircraft manufacturing (when factory worker housing was sometimes built in proximity to airports), relatively few residents of today's airport neighborhoods benefit individually from proximity to aviation industry jobs, or from ease of access to air travel. Geographic regions rarely directly reimburse localities for the costs of hosting airports, but airports sometimes work closely with one or more surrounding communities on projects of mutual interest.[61]

3.11 Aircraft Noise Effects on Property Values

Academic and commercial studies of real estate valuation have been conducted for decades. Most struggle to deal with the large number of amenities of properties that can affect sale prices, including house and lot sizes, numbers of bathrooms and bedrooms, age, type and quality of construction; neighborhood amenities (proximity to shopping, commuting time, quality of schools and other local government services, and the like); not to mention a wide range of negative externalities. Combinations of such amenities and externalities are typically idiosyncratic to neighborhoods, complicating the ability of researchers to control for all such factors when attempting to discern the effects of aircraft noise exposure in isolation.

Nonetheless, quantitative studies of the effects of aircraft noise exposure on residential studies are not uncommon. Early studies of this variety tended to be somewhat simplistic, in that they relied on the opinions of local realtors and appraisers who were, in some cases, well aware of the research hypotheses under test. Designs of studies conducted since the 1980s (when multi-variate statistical analysis software became widely available) tend to be more sophisticated than earlier studies, but still are not fully persuasive in isolating aircraft noise exposure effects.

In one study of military aircraft noise effects on housing values, Fidell et al. (1996) developed a multiple regression model of the sale prices of residential property in Pima County, AZ within the $L_{dn} = 65$ dB noise exposure contour of Davis-Monthan Air Force Base. Two sub-samples of these sales were drawn: one of 971 sales of 698 homes within the prior ten years, and one of 967 sales of 694 homes. The predictive model for property sale prices was developed from the first sub-sample of homes, validated in the second, and then applied to 14,326 residential property sales in Pima County that were *outside* of the air base's $L_{dn} = 65$ dB aircraft noise exposure contour. The same multiple regression model predicted property sale prices equally well both within and outside the air base's $L_{dn} = 65$ dB noise exposure contour. Noise exposure per se therefore had no measurable effect on sale prices in this case, after controlling for relevant differences in housing amenities county-wide.

The same multiple regression approach was then applied to an analysis of residential property sales in the vicinity of Langley Air Force Base in Hampton, Virginia. In this case, it was paradoxically found that average sale prices of residences *outside* of the air base's $L_{dn} = 65$ dB aircraft noise exposure contour were lower than those of residences within the $L_{dn} = 65$ dB noise exposure contour. The Air Force's prolonged attempts to discourage and delay construction of homes within its $L_{dn} = 65$ dB contour eventually led to the later construction of homes that were newer and larger than elsewhere in the city. The finding underscores the importance of real estate market-specific differences in factors that influence property values.

Nelson (1978, 2004) has nonetheless devised a generic "Noise Depreciation Index" (NDI) that purports to summarize the effects of aircraft noise exposure on residential property values in terms of a percentage discount per decibel of noise exposure, irrespective of local market effects. Nelson found from his 2004 meta-analysis that the approximate value of the NDI for residential property is about 0.6% per decibel of exposure. Thus, for example, Nelson concludes that all other things being equal, a property with a noise exposure of $L_{dn} = 75$ dB would sell for about 10 to 12% less than it would if its noise exposure were 20 dB lower ($L_{dn} = 55$ dB). In absolute terms, therefore, Nelson predicts that a $200,000 house would sell for $20,000 to $24,000 less if exposed to aircraft noise characterized by a level of $L_{dn} = 75$ dB rather than $L_{dn} = 55$ dB.

Desai and Chen (1994) investigated the feasibility of making a national-level determination of the effects of aircraft noise on property values. They concluded from their study of noise effects on housing values in three metropolitan areas (Baltimore, Los Angeles, and New York) that local conditions strongly affected property values, to such a degree that realtors and appraisers who were familiar with community-specific conditions were essential for such a determination. Their approach was to select pairs of neighborhoods with high and low aircraft noise exposure, controlling for covariates such as airport size, community socio-economic level, and airport employment, and to test hypotheses about the importance of each covariate with a series of statistical contrasts. With sufficient care and a large enough sample, they believed that a valid, national-level determination of an average estimate could in fact be derived. Their study was a methodological one that yielded no definitive conclusions about the actual effects of aircraft noise on property values.

3.12 Aircraft Noise Litigation

Communities sufficiently exercised by ongoing or anticipated aircraft noise exposure have been more successful in political and legislative than in legal challenges to the continued operation and expansion of airports. The usual legal challenges to aircraft noise exposure have included inverse condemnation (taking of property without due process of law) and nuisance actions, and procedural violations of NEPA-like (including state environmental impact disclosure) laws.

3.12.1 Small Claims Court Nuisance Suits

Conventional suits for recovery of aircraft noise-induced damages are such lengthy and expensive affairs that few individual plaintiffs bother to pursue them. As described by Freeman and Farris (1992), however, 172 plaintiffs filed claims in San Mateo County Small Claims Court in September of 1980 against San Francisco International Airport for the nuisance created by noise of aircraft departing SFO. The judge awarded the first round of plaintiffs an aggregate of $87,000.

Since the noise nuisance was a continuing one, plaintiffs undertook second and third rounds of Small Claims Court actions that ultimately yielded more than $300,000 in awards for noise damages. The first round of awards was reversed when the defendant appealed the Small Claims Court awards to Superior Court,[62] and SFO successfully lobbied the state legislature to preclude further small claims court noise nuisance suits. Most plaintiffs eventually accepted a settlement with the airport that they viewed as confirming that the airport took their grievances seriously. About two dozen plaintiffs who persisted in the later rounds of small claims court suits were ultimately paid their awards when the airport inadvertently failed to file timely appeals of their cases.

3.12.2 Conventional Inverse Condemnation Litigation

The cost, procedural complexity, and length of both individual and class action suits against airports for recovery of diminutions in property value lack the mass appeal of less formal nuisance suits, even though the monetary damages potentially available to plaintiffs in such suits can be far greater than those available from small claims court proceedings. The heyday of class action inverse condemnation litigation against airports appears to have passed, although both airports and land use zoning boards remain wary of them.

3.12.3 Challenges to Compliance with Environmental Impact Assessments

The National Environmental Policy Act (NEPA) and similar state-level legislation mandate disclosure of environmental impacts of major government-funded projects. Airport projects that receive federal funds must therefore comply with NEPA-like requirements. Such projects may meet NEPA requirements with a Categorical Exclusion (CATEX), Environmental Assessment (EA) or an Environmental Impact Statement (EIS). FAA's requirements for environment studies are delineated in FAA Order 1050.1F. Note that this document is an Order (as opposed to an FAA Advisory or Guideline), which makes the requirements mandatory for FAA.

It is beyond the scope of this monograph to describe the requirements for a CATEX, EA or EIS, but suffice it to say that a CATEX can only be granted for actions deemed eligible in the FAA Order. Even then, they may require some study to determine qualification for a CATEX. An EA is a more detailed analysis that generally focuses on specific issues. An EIS, however, must address all environment issues defined in FAA's Order, and generally requires a multi-year, multimillion dollar process. The EA and EIS processes require public disclosure and mandate public consultation programs. More than a dozen states have distinct environmental disclosure requirements, while others rely solely on the federal process. California (CEQA) and Minnesota (MEPA) legislation impose particularly stringent environmental review processes.[63]

Nearby communities commonly challenge the adequacy of the environmental assessment process for controversial airport infrastructure projects, usually on procedural grounds such as alleged failures to meet federal or state requirements or both. Nationwide, the most recent rounds of such litigation have concerned the implementation of FAA's new air traffic management programs, known as NextGen. NextGen litigation is of particular interest because the plaintiffs tend not to be communities nearest airports, but those miles away from runway ends that are overflown in new or revised flight corridors.

Litigation over the adequacy of the environmental process is complex, and can turn on procedural as well as substantive environmental issues.[64] Provided that all procedural requirements have been met (including meaningful notification and community engagement in NextGen planning), the results of such litigation tend to favor aviation interests. This is because the current national threshold of significance of aviation noise impacts ($L_{dn} = 65$ dB) is easily met in communities distant from airports. Even when the policy threshold is exceeded, the usual consequence is merely revision of the NEPA analysis, and re-publication of the environmental disclosure document, possibly incorporating additional mitigation measures.

Chapter 4
Aircraft Noise Measurement and Modeling

Community exposure to aircraft noise can be both measured and modeled. Modeling (i.e., prediction) of noise exposure is unavoidable in prospective applications, such as disclosure of the impacts of proposed airport infrastructure construction, since noise that has not yet been created cannot be directly measured. Measurement is more useful for retrospective purposes, such as monitoring and reporting of noise created by ongoing airport operations, and also for aircraft noise certification, per FAR Part 36.

Reliable measurement and modeling of aircraft noise exposure in airport environs took decades to achieve. Field measurements of aircraft noise exposure in the vacuum tube era required cumbersome and expensive equipment and specialized skills. Predictions of aircraft noise exposure through the era of mainframe computing likewise required complex, custom software. This chapter reviews the development of modern, standardized means for measuring and modeling aircraft noise exposure.

4.1 Development of Aircraft Noise Measurement

Attended, broadband (encompassing many frequencies) field measurements and spectral (sound level at each frequency) analyses of aircraft noise using portable equipment were highly time and labor-intensive prior to the development of digital instrumentation for measuring integrated noise exposure in the 1970s. Narrow band analyses of aircraft flyovers in the 1960s required rooms full of analog filter banks and tape loops, as well as other special-purpose equipment that was expensive to assemble, maintain, and operate. Field measurements were of limited dynamic range and duration, and difficult to coordinate with aircraft position information. Planning multi-site data collection exercises; calibrating, shipping, and installing equipment;

S. Fidell and V. Mestre, *A Guide To U.S. Aircraft Noise Regulatory Policy*,
https://doi.org/10.1007/978-3-030-39908-5_4

staffing data collection sites; not to mention post-processing of raw field measurements, all required a great deal of logistical effort and very high cost. Field measurements of aircraft noise therefore tended to be of modest scope, such as short duration, spot measurements at modest numbers of points.

4.1.1 Introduction of Digital Instrumentation

The digital sound level meters introduced in the 1970s had such a profound effect on aircraft noise field measurements that they quickly replaced analog instrumentation. The initial digital sound level meters, incorporating firmware-implemented fast Fourier frequency analysis (Cooley and Tukey 1965), reduced refrigerator-sized components to briefcase-size meters. With sufficient external batteries, longer term, unattended measurements of aircraft in the field became feasible. The first self-contained, portable, long-term (on the order several days to a week) monitors were introduced by BBN and Digital Acoustics in the mid-1970s. Shortly afterward, major vendors of acoustic instrumentation, such as Bruel and Kjaer and Larson-Davis, began producing similar meters. These systems were designed primarily for longer term spot measurements; that is, unattended measurements over several days or even weeks (albeit with periodic battery replacement and downloading).

Spot noise measurements, especially those done over several days or weeks, proved useful for defining 24-hour noise exposure and estimating long-term, cumulative noise exposure. However, such spot measurements were prone to sampling errors in estimating values of long-term exposure metrics around airports. Challenges to such measurements questioned whether airport operations during the sampling period were representative of annual average operations.

4.1.2 Aircraft Noise Certification for Purposes of FAR Part 36

Part 36 of the Federal Aviation Regulations requires that aircraft must be certified to meet approach, departure, and sideline noise-level limits. This regulation is harmonized with international regulations to create a uniform set of noise regulation worldwide governing the sale and operation of aircraft across international borders. The noise measurement for this certification process is highly controlled in terms of aircraft operation, limits on meteorological conditions for testing, acoustic instrumentation, aircraft position data collection, and sampling requirements. Initially, aircraft were required to meet the so-called Stage 2 noise limits. Those aircraft not certified to Stage 2 limits were deemed Stage 1 aircraft.

Aircraft entering service later had to meet Stage 3 noise limits, and eventually yet lower Stage 4 limits. Aircraft entering service today must meet Stage 5 limits. The only requirement to prohibit operation of older aircraft in the United States was for the phase out of Stage 2 aircraft over 75,000 pounds before January 1, 2000, due to the Airport Noise and Capacity Act of 1990 (Stage 2 aircraft under 75,000 pounds were phased out by the FAA Modernization and Reform Act of 2012). Measurements made for aircraft certification purposes are not conducted in a way useful to measuring aircraft noise in communities, so are not described further here. Noise certification data are, however, incorporated in the noise database used in aircraft noise modeling.

4.1.3 Representativeness of Field Measurements

Characterizing long-term, cumulative noise exposure at airports with multiple runways, at which runway use changes regularly with wind patterns, is particularly challenging. Was the weather during the sampling period representative of the annual average weather, particularly winds, temperature, and importantly, temperature inversions? Consultants formerly reported a few days of DNL measurement as a definitive measurement of annual average DNL. In 2007, however, the A-21 Committee on Aviation Noise and Emissions Measurement and Modeling of the Society of Automotive Engineers (SAE) published a guidance document on estimation of the uncertainty of measurement and proper sampling methods for the multiple airport operational modes over a period of a year (SAE 2007).

Short-term measurements are not likely to demonstrate with useful certainty whether noise policy guidelines are exceeded, nor to demonstrate the validity of noise modeling calculations, unless the measurement protocol is rigorous and of adequate duration. Such samples likely require many weeks of measurements covering all of the airport operating modes and parts of each season. International standards also specify performance requirements for sound level meters used for such purposes (IEC 61672, 2013).

4.1.4 Permanently Installed Aircraft Noise Monitoring Systems

Today it is common for airports to operate permanent noise monitoring systems. These were first introduced in the early 1970s and were minicomputer-based (PDP-8, General Data Nova for example). The State of California required any airport declared to have a "noise impact" to operate a noise monitoring system to help define the location of the 65 CNEL contour (California Administrative Code, 1971). As a result, major airports in California such as LAX, SFO, OAK, SNA, and SAN installed such systems. The first systems introduced included systems by EG&G, Tracor, and Olson Laboratories. Many airport noise monitoring systems vendors, including Bruel & Kjaer, Harris (formerly Tracor), Bridgenet, .01dB, Topsonic, and Casper, sell such systems today.

Airports that make intensive use of noise monitoring systems include those with pre-ANCA aircraft noise limits that penalize operators for exceeding specified noise limits set at monitors. These include John Wayne Airport, Long Beach Airport, and Santa Monica Airport. These airports also use data from noise monitoring systems for curfew enforcement. Denver International Airport relies on both noise modeling and measurement to enforce a pre-ANCA inter-governmental agreement between the airport proprietor and an adjacent county.

A permanent airport noise monitoring system consists of several components. Field-installed noise monitors consist of a sound level meter, microphone, and communications hardware and software. The microphone is installed on a mast at least 6 m high[65] and at least 3 m from any reflecting surface, such as a roof or building façade.[66] Data from the field monitors are sent to the central data collection system via a telephone lines, cell phone system, or the internet.

A second major component of noise monitoring systems is a means or collecting aircraft position data. It wasn't until the 1990s that FAA provided limited access to radar data. Today, FAA publishes aircraft position data via a system called System Wide Information Management (SWIM). SWIM collects data from traditional airport radar systems as well as much more modern position data such as multi-lateration, Automatic Dependent Surveillance-Broadcast (ADS-B), and modern ground radar Airport Surface Detection Equipment, Mode X (ASDE-X) systems. SWIM ranks the data by quality and publishes the best position data on a near real-time basis.

Aircraft position information is used in noise monitoring systems for two purposes. First, the position data is used to verify that noise recorded at a field monitor corresponds temporally to aircraft overflights, to distinguish aircraft noise from that emitted by ground noise sources. A second purpose is to support publication by airports of aircraft flight track maps as part of their noise management programs. Some airports (Seattle-Tacoma International Airport, for example) have fly quiet award programs identifying airlines based on the fleet average noise level and compliance with their recommended noise abatement corridors. Airport noise monitoring systems usually include a meteorological station or an internet connection to the airport's weather station. The system software is capable of producing a large variety of reports and graphics as well as internet displays of the data for public use.

4.1.5 Uncertainty of Aircraft Noise Monitoring

An airport's permanent noise monitoring system eliminates the uncertainty of measurement associated with the sampling error of spot monitoring. If the monitoring system is operating continuously, the entire population of flight operations is collected, so that no issue remains of whether the measurements captured a representative sample. Continuous monitoring does not eliminate the issue of uncertainty of measurement associated with a number of inherent uncertainties. The first of these is the measurement uncertainty of the sound level meter and its associated microphone. This is governed primarily by the IEC standards noted above for sound level meters.

Two international best practice documents prescribe the installation and operation of airport permanent noise monitoring systems. One is the aforementioned SAE Aerospace Recommended Practice (ARP) 4721 document, while the other is the International Standards Organization (ISO) 20906 (ISO 2009). The ISO document includes an Annex B that works through the uncertainty of measurement for Sound Exposure Level (SEL) measurement of a single flyover based on a measurement system that just meets the performance requirements of IEC 61672 (IEC 2013). SEL is the building block of DNL. The uncertainty of measurement of an SEL measurement for a Class 1 sound level meter (the most accurate of the Class ratings in IEC 61672) is ±0.8 dB. This level of imprecision yields a 95% certainty that the true SEL values lies within a value of the measured value −0.8 dB and the measured value +0.8 dB.

The State of California specifies that the required accuracy of a noise monitoring system that meets the requirement of California code must have an end-to-end accuracy of at least ±1.5 dB for the measurement of CNEL. Note that the California requirement is for a system designed to measure in the vicinity of the 65 dB CNEL contour. (For the purposes of the current discussion, CNEL and DNL may be considered nearly equivalent.) This is important as it is commonly thought that measurement data represent a "gold standard" for estimating true values of noise exposure. Although measurements may be a gold standard, they are not of absolute precision. Even the best of measurements are estimates that are subject to measurement uncertainty.

4.1.6 Implications of Uncertainty of Aircraft Noise Measurement

Measurement uncertainty is an important consideration in any discussion of aviation noise policy. Because L_{dn} is a long-term average, a number of factors contribute to measurement uncertainty of L_{dn}, including the instrumentation uncertainty, sampling uncertainty for non-permanent measurement installations, uncertainty due to atmospheric propagation effects, uncertainty due to varying meteorological conditions, uncertainty in aircraft operations (weight, flight procedures, flight paths, etc.), and uncertainty due to background noise effects. What is not commonly understood is that this estimate of uncertainty, such as ±1.5 dB at the 65 dB L_{dn} contour, is considerably smaller than the uncertainty at lower L_{dn} values. The location of the 65 dB L_{dn} contour can thus be specified with less uncertainty than the location of lower level contours such as the 60, 55, 50, or 45 dB L_{dn} contours. Other sources of uncertainty, primarily those associated with lower signal to noise ratios and the vagaries of long-range atmospheric propagation of sound, must also be considered at such lower levels.

The signal-to-noise ratio at lower aircraft noise levels is a major impediment to precise measurement. At an L_{dn} value of 65 dB, single-event aircraft noise is usually great enough that other noise sources in the built environment do not appreciably interfere with measurement of aircraft noise, as long as the measurement microphone is judiciously located (i.e., not adjacent to a major road or near air-conditioning units or any other competing noise sources). When the aircraft noise levels of concern are nearer to $L_{dn} = 55$ dB, competing noise sources can partially mask aircraft noise events, so that the uncertainty of measured aircraft noise level is greater than at higher single-event aircraft noise levels.

When ambient noise levels and aircraft noise levels are similar in magnitude, the issue is not simply how well a monitoring system measures noise, but how reliably it can differentiate aircraft from community noise sources. Total noise exposure has a lower uncertainty than estimates of either aircraft noise alone, or community noise alone.

Ambient noise levels in the built environment can be roughly estimated based on population density (Volpe 2010). Figure 15 shows the community DNL versus population density in persons per square mile. For example, the Volpe report shows data for many communities including San Francisco and some surrounding suburbs. The highest population densities on the San Francisco peninsula range from about 72,000 people per square mile in downtown San Francisco to about 10,000 people per square mile in San Bruno, a suburb south of San Francisco. Figure 4.1 predicts a L_{dn} value of about 58 dB for San Bruno, but an L_{dn} in excess of 65 dB in the heart of San Francisco. A community would require a population density of fewer than 4,000 people per square mile to attain a L_{dn} value lower than 55 dB. This provides only a general guideline for estimating whether community noise levels confound measuring aircraft noise. The L_{dn} generated by aircraft noise can be generated by a small number of relatively loud aircraft or a large number of relatively quieter aircraft. It is this latter case which is most common and the greatest potential for contamination by community noise sources.

The two concerns in areas with lower aircraft noise levels, say below $L_{dn} = 60$ dB, are (1) that some aircraft will produce noise at levels lower than those of community

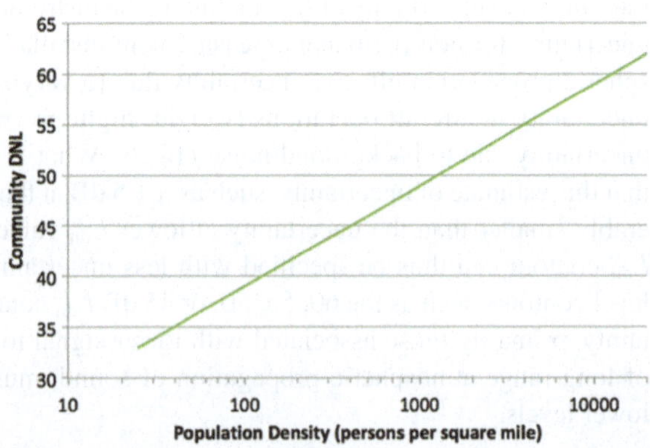

Fig. 4.1 Relationship between day–night average sound level and population density

noise sources, and (2) that some aircraft noise events will occur at the same time as community noise events such as motorcycles, trucks, buses, and power garden equipment. When aircraft noise at levels lower than community noise levels occurs, the measured DNL due to aircraft traffic will be under-estimated.

The degree of under-estimation may be small unless the number of missed flyovers is large. In the case of simultaneous aircraft/community noise events, DNL due to aircraft DNL will be over-estimated. Since noise monitors don't recognize noise sources as people can, such errors of estimate problems can contribute to the uncertainty of measured aircraft noise levels, particularly at sites distant from airports.

4.1.7 Classification of Monitored Noise Events

A noise monitoring system may record a noise event, and flight track information that may confirm the presence of an aircraft flyover at the same time, but current technology cannot determine what portion of a noise event is due to aircraft noise and what portion is due to other (community) sources. The human ear (and brain) may readily recognize the superposition of two noise events, but technologies for doing so in aircraft noise monitoring systems are still in limited use or under development. These include measures such as multiple microphone arrays to detect the direction of arrival of acoustic energy, which can help to distinguish airborne from groundborne noise sources.

Such measures would be useful at sites distant from the airport, but would not be helpful in cases where ground-based aircraft noise is an issue (e.g., aircraft maintenance engine runups, reverse thrust landing noise, and takeoff roll noise). Similarly, closely spaced (100'–500' apart) multiple noise monitors could compare the time of an event's maximum noise level at each monitor to distinguish between a nearby community noise source and a distant aircraft noise. These are not widely used techniques in current systems, but merit further study as interest in noise levels far from airports may become more widespread.

Figure 4.2 shows the capture rate for aircraft noise events as a function of distance of the aircraft from the microphone of an airport's noise monitoring system. Note that this figure is unique to a particular microphone site, aircraft fleet mix, and whether the microphone is under or to the side of the flight corridor. Typically, an airport noise monitor system will initially be configured to associate all flights within 1 nautical mile of the microphone to noise events. If the aircraft is a relatively quiet type (e.g., a small regional jet) at the outer limit of this distance, no noise event may be noted for that flight. "Tuning" a noise monitoring system to local conditions includes selecting the appropriate noise threshold trigger, minimum noise event duration, and distance to aircraft for noise event matching. This is a critical step in the operation of an airport noise monitoring system and may need to be repeated as the aircraft fleet changes or as flight corridors are modified or condensed.

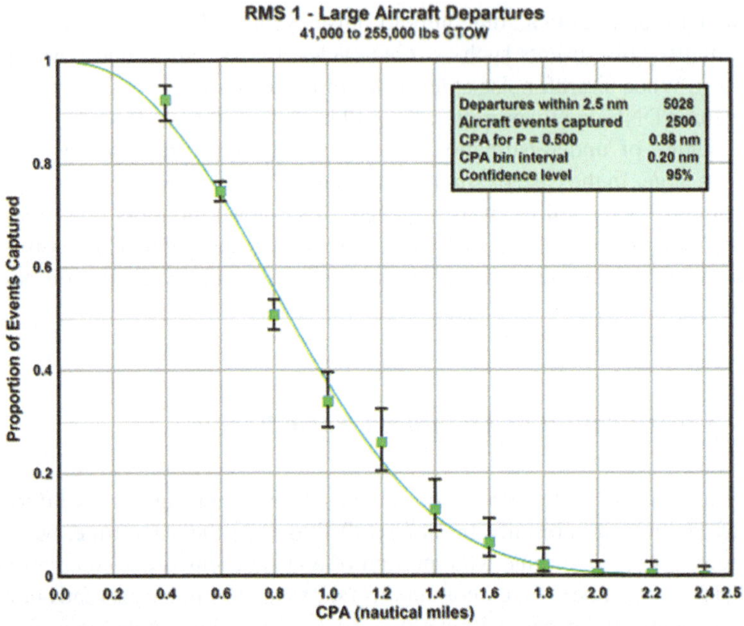

Fig. 4.2 Capture rate of aircraft noise events as a function of closest point of approach (CPA) by one conventional aircraft noise and operations monitoring system (Fidell et al. 2001)

Figure 4.2 shows a decrease in probability of noise event capture as a function of distance. This is easily misconstrued as a failure to accurately measuring cumulative noise exposure. Unless extreme precision of measurement is at issue, measuring a cumulative noise metric like DNL requires only the capture of the high-level noise events. Low-level noise events, unless extremely numerous, have only minor effects on estimates of cumulative noise exposure.[67]

Another factor that may affect the uncertainty of short-term measurements is varying meteorology and its effect on sound propagation. For measurements around or near the 65 DNL contour in most cases are close enough to the aircraft that these effects may be small. But when measuring at lower level DNL values the effect of wind and temperature gradients can have large effects on the measured noise levels. Ensuring that a sample period includes representative temperature gradients is very difficult. In a normal atmosphere temperature decreases with altitude. When there is a temperature inversion, increasing temperature with altitude, sound waves are bent back toward the ground and noise levels in some locations will be higher than for normal (adiabatic) conditions. Temperature inversions are common at night and may extend into the late morning hours, especially during winter months. This means that to estimate annual DNL the best practice may need to include measurements from different seasons. That may vary from place to place in the country as some areas, such as Southern California, effectively have only two seasons. The Midwest and East Coast may require more sample periods to cover the various propagation conditions.

Specialized monitoring is subject to different sets of concerns. It is common now to require indoor–outdoor measurements to determine the sound transmission of outdoor noise levels to the indoors for school and residential sound insulation programs. It is beyond the scope of this monograph to address these tests in any detail. In these tests a microphone system is placed outdoors and another indoors and the difference is measured and reported as the transmission loss. These tests may use actual aircraft noise or a speaker system to simulate aircraft noise. Because these tests are measuring a difference across the building façade, meteorology plays no role and such tests can be done quickly. The American Society of Testing and Materials (ASTM) has a measurement standard for this purpose (ASTM 2018). The ASTM standard covers many different sources and many different measurement situations and includes multiple methods. As of this writing, new guidelines for best practice specifically for aircraft indoor–outdoor measurements are under development by the FAA and SAE.

4.1.8 Smartphone Noise Measurements

Smartphones have become powerful tools for data acquisition and processing by sophisticated audio software packages. Questions have been raised about use of a smartphone application program to measure aircraft noise. A typical smartphone, using its internal microphone, does not support such uses, since the internal microphone of a smartphone is inadequate for measuring environmental noise. Smartphones lack readily available microphone windscreens, as well as means to acoustically calibrate the system. All outdoor measurements must include a windscreen, as even a light breeze can corrupt sound level readings. The inexpensive microphones in smartphones also fail to meet standard precision requirements for acoustic measurements.

A Class 1 sound level meter requires an expensive precision microphone. Attaching a Class 2 rated omni-directional microphone to a smartphone can make for a reasonable measurement system. Chucri et al. (2016) discuss this matter in detail. The question of whether a more precise Class 1 system or a Class 2 system as defined by IEC 61672 is worth revisiting. The specifications for the Class 1 and Class 2 system make it clear that a Class 2 system is not very good for measuring low frequency noise, as for example, that of helicopters. But for measuring commercial jet operations, with broadband noise signatures, Class 2 measurements may yield reasonable estimates of local noise levels. Sampling requirements and proper microphone locations are equally important considerations.

4.1.9 Development of Aircraft Noise Modeling

Manual methods were used to prepare airport noise exposure contours through the 1960s. Development of computer-based aircraft noise modeling has since followed a decades-long evolutionary path that has yielded two major software systems, variants of which are in use worldwide. The approach used to model aircraft noise is to construct a large grid of points on the ground that covers the desired modeling area around an airport. Noise levels are computed at each grid point, and a contour mapping program draws contours or isopleths (lines of equal noise value) through the grid point calculations. The mathematics of computing the noise level at each grid points is complex. It includes calculating slant range distances from aircraft to grid points, computing speed, altitude, thrust, and reference noise emission level at many points on aircraft flight paths, and using sound propagation algorithms to predict the noise exposure produced by each aircraft. The process is repeated for each aircraft operation on an annual average day.

4.1.10 Noisemap

The U.S. Air Force's NOISEMAP software was first developed in the late 1960s to improve on hand-drawn noise exposure contours.[68] Subsequent Air Force and FAA noise modeling software evolved largely from this early effort. All of these models predict cumulative, long-term average noise exposure. Although short-term single-event noise can be computed, they are not intended to replicate the conditions of specific flights.

There are few substantive differences today in the approach to modeling used by the military and by the FAA. Perhaps the most significant difference is that the FAA models contain a database of aircraft departure and arrival procedures for common aircraft types. NOISEMAP does not have such standard profiles, but requires the user to consult with each military squadron to define their unique procedures and develop custom profiles for each squadron. This approach was adopted because of the many variations in training and mission protocols used for a large variety of aircraft from trainers, fighters, bombers, and heavy lift helicopters. For purposes of this monograph, the discussion focuses on civil aircraft noise modeling conducted with FAA software.

4.1.11 FAA Integrated Noise Model (INM)

FAA introduced the first version of its Integrated Noise Model (INM) in 1978. The initial version of INM, which ran on a mainframe computer, was very cumbersome to use. The mainframe version of the software was ported to the personal computer in 1982, greatly facilitating airport noise modeling. INM was the workhorse of FAA noise modeling through version 7.0d, released in May of 2013. Many improvements and enhancements to INM were made between the time of its initial release and the time that it was supplanted by AEDT (see Sect. 4.1.12) in May of 2015. These included updated aircraft noise and performance databases, improved atmospheric propagation modeling, engine mounting effects, and many others.

FAA does not modify its noise models based solely on internal considerations. Industry best practice guidance was introduced in 1985 with the publication of SAE Aerospace Information Report (AIR) 1845 (SAE 1985). The European Civil Aviation Conference (ECAC) also publishes detailed best practice guidelines for aviation noise modeling. In 2008, ICAO published a document on best noise modeling practice (ICAO Doc 9911, 2008). Since not all of ICAO's national representatives have relevant technical expertise, the technical components of ICAO's best practice guidance were developed by the SAE A-21 Committee on airport noise and emissions modeling, and by the European Civil Aviation Conference (ECAC). ECAC published Document 29, "Report on Standard Method of Computing Noise Contours around Civil Airports", which includes the latest updates to modeling including many updates to SAE AIR 1845.

4.1.12 FAA Aviation Environmental Design Tool (AEDT)

FAA's AEDT, first released for use at US airports in May 2015, has fully supplanted INM. AEDT permits modeling not only of noise modeling but also of air emissions, dispersion and fuel burn modeling in a single software system. AEDT can model noise exposure produced by multiple airports, allowing the modeling of noise created by an entire airport system, as well as comparisons of alternative airport development or operational scenarios.

AEDT permits users to directly specify aircraft flight profiles, and to compute them from flight procedures. For example, a user can specify speed, altitude, and engine power setting at various points along the flight path, or to describe the flight profile in terms of a procedure such as takeoff roll, climb step, acceleration step, and combined climb/acceleration step throughout the flight path. If the latter approach is adopted, AEDT can compute altitude, speed, and power setting (thrust), including the effects of air temperature and winds on the flight profile. Composing a noise model for a busy airport requires a great deal of operational data, including the airport layout, digital terrain databases for the airport environs, numbers of operations by aircraft type by runway, and stage length and flight track by time of day. AEDT can also import automated digital information about aircraft position reports.

AEDT includes a library of common approach and departure profiles for commercial and general aviation aircraft. A key to accurate modeling is having the aircraft in the correct location at the correct thrust for each segment of the flight. Distance and thrust are the two most important variables, neither of which is provided by the user. These are, instead, computed from the either the standard profiles or user-provided profiles. The FAA models, INM and AEDT, include alternative profiles based on ICAO noise abatement procedures, as well as unique profiles for aircraft weights varying from light short flights to heavy long-haul flights.

ICAO's goal is for its member nations to harmonize noise modeling according to Document 9911, and to incorporate the best practices developed around the world. AEDT (see Sect. 4.1.12) is kept Document 9911 compliant by the FAA. As part of this international effort to harmonize aircraft noise models, the underlying aircraft noise database and aircraft performance data are kept in mutually available database for all users domestic and international, including AEDT users. These centralized databases are maintained by Eurocontrol (a pan-European organization supporting European aviation) for all users.

Aircraft Noise Model Accuracy

The accuracy of noise modeling is not well established but there is no doubt the evolution of noise modeling has greatly improved the reliability of the models. There are many sources of uncertainty in noise modeling. These include uncertainty in knowing the number of operations by aircraft type by time of day, the distribution of aircraft weights, the flight profiles (speed, altitude and thrust), runway use, flight track assignments, and the lateral and vertical distribution of flight tracks. In addition, there is the uncertainty of the noise–power–distance curves (the database of noise by thrust setting as a function of distance from the aircraft), average airport temperature, and wind conditions. Selecting an average airport temperature requires the modeler to decide whether to use average 24-h temperature, daylight average temperature, or nighttime average temperature, based on the distribution of the dominant airport operations.

As described in earlier paragraphs, two key factors to model accuracy are knowing the correct aircraft position in space and modeling the correct thrust for each point along the flight path. For estimating the location of the 65 dB L_{dn} contour (the primary use of aircraft noise models today), the distance between an aircraft a point of interest on the ground is short enough that long distance sound propagation variations are not a major issue. The models account for the much greater atmospheric absorption of high-frequency than low-frequency sound. They assume a homogeneous atmosphere, however: one through which sound propagates under uniform temperature, humidity, wind speed, and wind direction conditions. Modelers can specify each of these, as well as a corresponding wind speed for each runway end, but the specified values are applied to the entire atmosphere between the aircraft source and points on the ground.

In reality, each of these atmospheric variables may vary from point to point. The disparity between reality and modeling assumptions does not pose a major issue over short propagation distances. When attempting to model noise levels much lower than $L_{dn} = 65$ dB, the homogeneous atmosphere assumption is weak, so the resulting uncertainty of the lower level modeled DNL values is much greater than at higher DNL values. The real atmosphere includes temperature and wind gradients and wind shifts. Over long distances, these gradients can alter sound levels by focusing sound, much as a lens focuses light. While it is possible to create models of propagation through temperature and wind gradients, the actual temperature and wind gradients in a real atmosphere cannot be known over long distances, nor on an average annual day basis. Any discussion of noise policy must consider that higher level DNL values can be modeled with much less uncertainty than lower level DNL values.

FAA has been studying aircraft noise modeling uncertainty since at least 1978. In 1982, FAA published a comprehensive study of measurement versus modeling uncertainty (FAA 1982). The study included a comprehensive comparison of measured and modeled noise levels at Seattle-Tacoma International Airport. This study compared predicted Sound Exposure Level (SEL, the building block of DNL) to measured values for generic aircraft categories. The range of differences ranged from fractions of a decibel to +6 and −7 dB. The mean difference was 3 dB. All of the sites used for this study were within 4 nautical miles of runway ends, so the study pertains directly only to the uncertainty of the higher DNL values.

The uncertainty of 3 dB in SEL values is only one component of the uncertainties of modeled DNL, however. Additional uncertainty arises from uncertainties about numbers of operations, time of day of flight, and most importantly, variation in the vertical and horizontal flight path profiles. These 1982 findings probably led to the following statement made in FAA's 1983 Advisory Circular 150/5020-1 (the implementing guideline for Part 150 of the Federal Aviation Regulations):

> "**233. ACCURACY.** As is the case with any computer program or with any prediction method, the accuracy of the output of the Integrated Noise Model is directly dependent upon the appropriateness, completeness, and accuracy of the input data. Use as input of average flight tracks, flight procedures, aircraft types and mix, and the schedule of operations can degrade the accuracy of the predicted contours. Further, the effects of local topography, weather, buildings, etc., cause variations from point to point along a contour. Accordingly, the accuracy of the INM computer noise prediction model in estimating the yearly average L_{dn}, value at any specific geographical point has been estimated to be L_{dn} 75 contours ±3 dB and L_{dn} 65 contours ±5 dB with the average error over all points along the contour tending towards zero".

This statement of accuracy made by the FAA in 1983 has not been updated in such an explicit form since 1983. No doubt modeling methods and database improvements have reduced the uncertainty, but there has been no similarly explicit statement published. In 2017, FAA published a comprehensive guide to the uncertainty of the AEDT model that exhaustively looks at each variable and its effect on uncertainty (FAA 2017). The 2017 report makes no attempt to work out an overall measure of uncertainty in units of DNL, as such an overall estimate of uncertainty requires knowledge of the uncertainty of each individual variable, and of methods that may vary from case to case.

A similar examination of modeling uncertainty undertaken in Europe (Eurocontrol 2002) addressed the uncertainty of individual variables and methods, but made no attempt to combine these individual uncertainties into an overall DNL uncertainty. Although it is not possible to generalize an estimate of uncertainty for modeled DNL in all cases, uncertainty limits can be evaluated in specific cases. Anecdotally, most modelers consider that ±1.5 dB of uncertainty is achievable in the vicinity of the

$L_{dn} = 65$ dB when good operations and flight track data are available for a sea-level airport lacking extraordinary terrain or meteorological anomalies. Uncertainty would increase at lower DNL values. This assessment is based on casual comparisons of noise model results at airports with permanent noise monitoring systems. A large survey of such airports and their noise modeling would be of great value.

Contour Mapping

The density of the analysis grid determines the resolution of the contours. A coarse grid, with large grid point spacing, will produce a rougher edged contour than a finer grid with less distance between grid points. In fact, the contouring software packages contain parameter settings that will draw slightly different contours from the same grid points. The implication of the mapping interpolation is an added uncertainty of model estimates (discussed earlier) that needs to be understood as part of policy development.

Overall public understanding of aircraft noise exposure contours is often poor. The public frequently misunderstands the unit of long-term cumulative noise exposure (DNL) that is most often contoured, and misinterprets the "annual average day" condition modeled as a noise condition that can actually be heard. Some airports have pronounced seasonal shifts in operating patterns. Vacation destinations, for example, often have strong summer or winter seasonal peaks in flight schedules. Modeling noise exposure at such airports on an average annual day basis risks misrepresenting the aircraft noise that drives actual community response.

Comparing community noise complaint patterns to long-term noise exposure contours is thus problematic. A complaint, while possibly influenced by long-term exposure, is often linked to one or more specific noise events. These can include anomalous flight operations that are so infrequent as to exert negligible influence on annual average noise exposure contours. Community attitude surveys, on the other hand, are designed to quantify long-term, community-wide attitudes about aircraft noise. Community attitude surveys, rather than complainant locations, are used to construct relationships between long-term noise exposure and community response.

Misunderstandings of noise exposure contours, especially when rendered as two-dimensional isopleths at 5 dB intervals, are also common. The public sometimes misinterprets noise exposure contours as marking staircase-like changes in noise levels. Such presentations can be misleading. Noise varies continuously, not in discrete intervals. Thus, a resident on one side of a noise contour that defines, for example, a hard boundary of areas eligible for sound insulation within which airports will pay for sound insulating windows and doors, will have an imperceptible difference in sound level from that of a neighbor whose home lies on the opposite side of the defining contour.

Using such "hard line" boundaries to exclude portions of communities from residential acoustic insulation programs can contribute to airport noise controversies. While planning practice may say such effects are inevitable, they contribute as a non-acoustical factor that may result in higher adverse response to the noise than would be expected based on the acoustic exposure alone. The use of these hard insulation boundaries has the opposite effect than what was expected in terms of community benefit, that is, while the beneficiaries of home insulation are generally satisfied by the results there is a new community of rejected homeowners that become displeased with the airport and may lead to claims of misfeasance or malfeasance by airport management.

4.2 Reconciling Aircraft Noise Monitoring and Modeling

Noise measurement data is generally considered the "gold standard" in defining the noise at a given location. Where noise measurements show a different noise level than noise modeling, it is common to assume the measurement data are a better estimate of noise exposure or to use such data as a claim that the noise modeling is not correct. That is not necessarily the case. Uncertainties of both aircraft noise measurement and modeling are unavoidable and comparing the two needs to recognize the uncertainties of each. For example, does the noise model estimate lie within the uncertainty of measurement? If so, then statistically the noise measurement data validates the noise model result and is not proof of a modeling error. For example, if the noise model estimates the DNL value at a location as 64.9 dB and measurements show the DNL value at 65.4 dB $\pm$ 2dB, then it can be concluded that the true DNL value lies between 63.4 and 67.4 dB. The noise model value is within that range, but nothing more can be known about the noise level at that location. This example also shows the absurdity of reporting such noise levels to tenths of a decibel. While policymakers want to know an exact DNL value, the reality is that policy needs to consider the underlying uncertainty of measurement and modeling. At lower DNL values, certainly at or below 50 dB, or below ambient noise levels as shown in Figure 15, noise measurement data may be of little use in validating noise modeling data.

4.3 Implications of Uncertainty for Regulation of Aircraft Noise Exposure

As noted by Fidell and Schomer (2006), measurements of sound pressure levels to a precision greater than a few tenths of a decibel are unattainable outside of a standards laboratory, while field measurements of acoustical quantities to a precision greater than the nearest decibel are prohibitively costly and generally impractical. Regulatory criteria are nonetheless interpreted as though they were the product of engineering calculations of unlimited precision. Thus, a measured or modeled aircraft noise exposure level of 65.1 dB is viewed as a requirement to disclose noise impacts.

It is therefore not surprising that thresholds of regulatory concern continue to be defined even today, with permanently installed noise measurement instrumentation and highly refined software, in 5 dB increments. (Recall further that these interpretive thresholds are of hypothetical "annual average day" cumulative noise exposure, not the noise exposure to which airport-vicinity residents may actually be exposed from day to day.) The practice of defining interpretive criteria in 5 dB increments arose originally from technical necessity, as a consequence of the limited duration of early field monitoring of aircraft noise exposure, and of the limited precision of early prospective software modeling of airport noise. The practice persists for lack of credibility of claims to any finer resolution of noise exposure estimation.

Chapter 5
Airport-Vicinity Land Use Planning

Aircraft noise is such a central concern in airport-vicinity land use planning that in practice, "land use compatibility" is a catch-all term for maintaining a buffer zone around airports in which land uses which pose a risk to the continued operation and expansion of airports are excluded from high noise areas. "Compatibility" is very much in the eye of the beholder, however. Although airports view nearby residential land uses as a risk to their continued operation and expansion, communities see airports as a threat to neighborhood quality of life, and possibly to municipal tax bases. Because the term "compatibility" usually implies some form of mutuality or reciprocity, land uses considered by airports as appropriate are referred to rather oddly (from a community perspective) as "compatible" ones.

5.1 Encroachment

Not even the most forward-thinking land use planning can guarantee long-term compatibility between airports and communities. "Encroachment" is the term that both airports and communities use to describe the growth of each toward the other. Both airports and communities have powerful growth incentives, and often encroach on one another. The reasons for such encroachment are simple to understand. Airports seek growth for a range of reasons: typically to meet existing or anticipated demand for air transportation services, but incidentally for the side benefits of increased aviation and sales tax revenue for their proprietors, and occasionally even to support local political interests.[69] Landowners and developers seek the most profitable uses of their land. Airports are intentionally located in centers of air transportation demand,

S. Fidell and V. Mestre, *A Guide To U.S. Aircraft Noise Regulatory Policy*,
https://doi.org/10.1007/978-3-030-39908-5_5

and can be a source of jobs, and consequently a magnet for residential development. Incompatible residential housing around some airports (as, e.g., in Long Beach, Santa Monica, Los Angeles, and Seattle) was in fact housing built to support jobs located at the airport. In all of these cases, the housing was built prior to the introduction of jet operations.

As described in Sect. 3.10, airports frequently describe themselves as "engines of economic development", and boast of their substantial contributions to local and regional economies. Oddly, they are often dismayed when such engines actually attract the economic activity that supports community growth. For their part, some community organizations may initially welcome airport growth plans for the commercial development opportunities that promise to increase sales and property tax bases, but later bemoan the aircraft noise exposure that can accompany such development.

From an airport perspective, local zoning authorities may permit or even encourage airport-incompatible real estate development in airport environs. Airports that were once surrounded by agricultural fields may eventually find themselves encircled by residential and other incompatible land uses. When they do, they invariably attempt to wring the last few percent in growth, even at excessive cost, out of their existing sites to accommodate increased demand, before even considering regionally-appropriate solutions, such as adding capacity at nearby airports or building new airports.[70]

Likewise, communities also have strong growth incentives. Residents' appetites for local government and utility services (e.g., police, fire, schools, sanitation, water and power, libraries, recreation, public health and transportation, and so forth) are nearly boundless. Such services are supported by combinations of municipal property taxes and fees paid to private sector service providers, and in many cases, may be facilitated by greater absolute population sizes and higher population densities, among other factors. Like airports, communities also compete with one another for growth. Communities with larger populations and tax bases can afford to offer residents more and higher quality government and private services. In short, both airports and some segments of communities typically view growth as their inevitable destinies. In good economic times, airport and community growth aspirations may converge geographically.

From the perspective of airport-vicinity communities, airports present a constant threat of unacceptable growth of noise exposure. Aircraft types serving the airport continually change, and not always for the better in terms of aircraft noise. While the long-term trend is to quieter aircraft, the fleet may change to larger heavier aircraft as demand changes, or to quieter narrow body aircraft. The long-term trend is for numbers of operations to continue to increase (subject to short-term decreases during economic downturns or industry-disrupting events) as long as regional demands for air transportation services continually increase. Airlines understand that they can often continue to make money in some markets by scheduling more convenient flight service, notwithstanding routine flight delays that accompany over-scheduling at airports with insufficient runway capacity. Conversely, at small general aviation airports with propeller-driven aircraft, the introduction of jet aircraft is not to a quieter aircraft but to a noisier one. This is a case where commercial airports and general aviation may differ significantly.

Airports' business models strongly incentivize growth, even when under-utilized nearby airports have unused, excess capacity. Airports earn revenue on a piece-rate basis: the busier the airport, the greater its revenue.[71] They collect landing fees on aircraft, head taxes ("passenger facility charges") on travelers, fees on parked and rented cars, leases on gates, and contract rent on fixed-base operators and in-terminal retail business space. Airports compete with one another both locally and regionally for flight operations, sometimes even subsidizing passenger gates and cargo handling facilities to attract new business.

If an airport has enough room, its first growth choice is generally another runway. Runway construction is a particularly attractive choice for airports because it can be heavily subsidized by federal grants. (Other airport infrastructure construction is more likely to require additional municipal bond and/or private financing.) Individual airports' short-term interests in squeezing the last few percent in capacity out of an existing facility, rather than acquiescing to the growth of nearby competing airports, can sometimes conflict with regulatory interests in providing efficient air transportation services for regional populations.

Figure 5.1 summarizes the growth in U.S. commercial civil aircraft (passenger plus cargo) operations from 1950 to 2017. During these nearly seven decades, total operations have added roughly a quarter million flights per year, and have grown from less than five million per year to nearly 25 million per year, at an annual growth rate of about 4.5%.

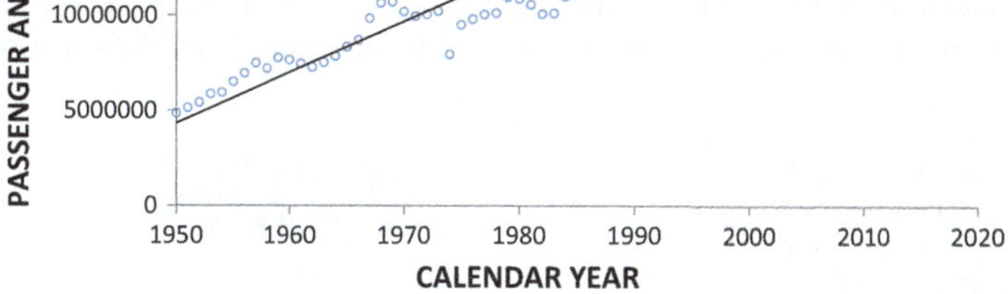

Fig. 5.1 Growth in annual passenger and cargo commercial aircraft operations at U.S. airports (from http://airlines.org/dataset/annual-results-in-u-s-airlines-2/)

5.2 ICAO's "Balanced Approach"

Land use compatibility is one of the four elements of the International Civil Aviation Organizations (ICAO) "balanced approach",[72] illustrated in Fig. 5.2. For U.S. domestic purposes, FAA has published its own recommendations for land uses near airports, pursuant to the Aviation Safety and Noise Abatement Act of 1979, Federal Aviation Regulation (FAR) Part 150, "Airport Noise Compatibility Program" (FAA 1984).

Limiting airport-vicinity land uses to those that do not pose risks to the continued operation and expansion of airports protects the public's investment in airport infrastructure, but is fraught with difficulties and limitations. Historically, land use controls have enjoyed only limited success, particularly in the vicinity of major urban airports. Many of the larger airports in the United States were already surrounded by residential land uses at the time of the introduction of jet air transport.

For land use planning purposes, airports can be classified into three general types:

(1) Those surrounded by residential land use at the time of jet introduction;
(2) Those surrounded by undeveloped or otherwise open space or compatible uses at the time of jet introduction; and
(3) Airports still surrounded by a mix of residential and open space.

The following three sub-sections address land use planning for these three categories in turn.

5.2.1 *Airports Surrounded by Incompatible Land Use*

Airports already surrounded by residential land use are common in higher population density and coastal regions of the United States. For such airports, land use compatibility programs provide either little or no relief, nor do they prevent adverse noise impacts. ICAO's "balanced approach" is clearly inapplicable at such airports. There can be no effective land use controls after the land is already incompatibly

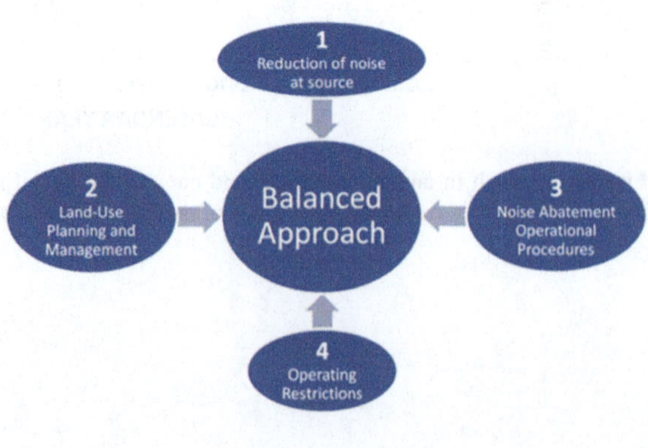

Fig. 5.2 The four elements of ICAO's "Balanced Approach" to aircraft noise management, per ICAO Document 9829

developed short of extensive residential purchase programs. These are discussed in a later section.

5.2.2 Airports Surrounded by Mixed Residential, Open Space, and Noise-Compatible Land

For airports that had both residential uses and open space land uses in the airport vicinity at the time of the introduction of jets, or even today, the problem of developing a noise policy that fits both land use patterns has been a key impediment to noise policy development. The policy conundrum of the 1970s that stifled adoption of progressive noise policies was how to design a noise policy that would prevent undeveloped land near airports from being developed for residential use, while at the same time *not* implicating existing residential uses at the same noise exposure levels as unacceptable (with a concomitant liability for such noise). The assumed cost of such liability has continued to pose an ongoing constraint on noise policy development, even though airports have largely avoided significant liability as a result of noise impact claims.

A detailed discussion of legal liability for aircraft noise impacts is beyond the scope of this monograph. However, as described in Sect. 2.5.3 and by Bennett (1982) and ACRP (2011), the Federal Government and the airlines are immune from such noise liability claims in the United States. The concern about noise liability dominated policy discussion in the 1960s and 1970s, and persists today. An early reference to potential noise damages may be found in the last section of the 1965 Housing and Urban Development Act. This legislation provided for, among other things, financial guarantees for housing both to individuals and to developers for the purpose of financing housing developments. The last section of this legislation provides the following requirement for HUD:

> SEC. 1113. The Housing and Home Finance Administrator shall undertake a study to determine feasible methods of reducing the economic loss and hardship suffered by homeowners as the result of the depreciation in the value of their properties following the construction of airports in the vicinity of their homes, including a study of feasible methods of insulating such homes from the noise of aircraft. Findings and recommendations resulting from such study shall be reported to the President for transmission to the Congress at the earliest practicable date, but in no event later than one year after the date of the enactment of this Act (August 10, 1965).

This is the first reference to effects of civil aircraft noise in the United States. In the context of the legislation the purpose was not altruistic but to provide assurance the mortgages backed by the Federal Government would retain their value, and not be damaged by loss of economic value due to aircraft noise. The legislation does not provide a reference to any study or documentation of aircraft noise impact on home value, nor did a literature search provide any such documentation as of the mid-1960s. The economic impact of aircraft noise appeared self-evident circa 1965, while later studies (*cf.* Sect. 3.11) have reached mixed conclusions on this topic, ranging from no effect to substantial effects (ACRP 2008).

Although the concern for financial liability has influenced noise policy making for decades, in the nearly 70 years of jet air transport significant liability has not been incurred by the FAA, airlines, or airports. This observation is not meant to denigrate the value of land acquisition and home buyout programs undertaken decades ago, as these buyouts occurred at noise levels well above the noise/land use policy level as described in FAR Part 150.

5.2.3 *Airports Surrounded by Undeveloped Land*

The last class of airports noted above are those with largely undeveloped surrounding land at the time of jet introduction. It is these airports for which efforts to achieve land use compatibility, ICAO's "balanced approach", and FAR Part 150 planning are likely to be most effective. These airports, many of which are in the middle of the United States, receive little attention because they experience few noise-related problems. Undeveloped land surrounding such airports may be merely accidental, however, and not the result of compatible land use controls.

5.3 Federal Noise/Land Use Compatibility Guidelines

The basic element in land use planning is definition of a noise level for which a given land use is compatible. (This is often simplistically called the "noise policy".) In reality, noise policy is a set of programs, regulations, and guidance for managing airport noise. The tendency to focus on the definition of that noise level that is deemed compatible for residential land use—currently $L_{dn} = 65$ dB – nonetheless persists. While the Federal Government has virtually no role in land use planning, the earliest guidelines for local government to use for land use planning guidance came from the Department of Housing and Urban Development (HUD 1971). The HUD guidelines were intended to prevent federally-backed loans for homes in high noise areas, and had no implications or effects on land planning that did not involve federally-backed mortgages. In the 1960s and early 1970s, most home purchases involved federally-backed loans. By the mid-1970s, home appreciation reached levels that exceeded the limits of federal loans and the HUD noise limits became essentially useless as noise control measure. Interestingly, mortgage lenders today have little or no direct concern with noise exposure as part of the lending process.

The FAA published noise/land use compatibility guidelines pursuant to the Aviation Safety and Noise Abatement Act in Federal Aviation Regulation Part 150, and in Aviation Advisory Circular 150/5020-1. Appendix 1 of AC 150/5020-1 includes a "Table of Land Uses Normally Compatible with Various Noise Levels". This table includes a category for residential uses other than mobile homes and transient lodging. The guidelines for residential exposures less than $L_{dn} = 65$ dB is "Y", 65 to 70 DNL and 70 to 75 DNL is N(1). The "Y" means "YES. Land Use and related structures are compatible without restrictions", while N(1) means "NO, Land Use and related structures are not compatible and should be prohibited". Footnote 1 of the Part 150 Table states that

> "Where the community determines that residential or school uses must be allowed, measures to achieve outdoor to indoor Noise Level Reduction (NLR) of at least 25 dB and 30 dB should be incorporated into building codes and be considered in individual approvals. Normal residential construction can be expected to provide a NLR of 20 dB, thus, the reduction requirements are often stated as 5, 10 or 15 dB over standard construction and normally assume mechanical ventilation and closed windows year round. However, the use of NLR criteria will not eliminate outdoor noise problems."

The meaning of the recommendations for residential exposures between DNL values from 65 to 70 dB, and between DNL values from 70 to 75 dB, is that it is generally not recommended. However, where local determination provides for such uses, the residences should have sound insulation such that outdoor-to-indoor noise reduction is 25 and 30 dB, respectively for the ranges described. In summary, for noise exposures up to $L_{dn} = 75$ dB, residential uses are not prohibited by these guidelines, but are discouraged unless adequate sound insulation is provided. Note that the FAR Part 150 Guidelines are written in a way that implies application to new construction. While the guidelines are often used in determination of noise mitigation measures, including eligibility, the treatment of existing homes and schools is subject to separate guidance from the FAA.

The FAR Part 150 noise/land use compatibility guidelines are neither a standard of performance nor a requirement. The footnote to the Part 150 Noise/Land Use Compatibility Guidelines emphasizes that:

> "The designations contained in this table do not constitute a Federal determination that any use of land covered by the program is acceptable or unacceptable under Federal, State, or local law. The responsibility for determining the acceptable and permissible land uses and the relationship between specific properties and specific noise contours rests with the local authorities. FAA determinations under Part 150 are not intended to substitute federally determined land uses for those determined to be appropriate by local authorities in response to locally determined needs and values in achieving noise compatible land uses."

FAA publishes a number of guideline documents for local governments and airports, in the hope that local initiative will implement some or all of these programs (FAA 2019). No matter what land use recommendations FAA publishes, no pre-emptive federal authority precludes any form of land use development in the vicinity of airports. The Federal Government simply has no control over land it does not own. Failures of airport-vicinity land use planning therefore may not reasonably be attributable exclusively to federal policy. Landowners have the rights to the highest and best use of their land, even when it is exposed to aircraft noise. National noise policy provides no incentives for local authorities to plan land uses for compatibly with airport operations, nor for builders to build compatibly.

5.4 Land Use Control Options Available to State/Local Government

The FAA and airports have long encouraged compatible land use development around airports, but it is only municipalities that have the authority to implement land use control measures. In some cases, the airport proprietor may be a city or county that has authority over some of the property around the airport. For example, the City of Los Angeles operates Los Angeles International Airport, and also has land use authority over those lands near the airport within the City of Los Angeles municipal boundary. Other lands adjacent to the airport fall within the jurisdictions of the County of Los Angeles, the City of Inglewood, and the City of El Segundo, however. Even when the airport proprietor has land use planning authority, development pressures on city hall and zoning boards from landowners and developers may outweigh the recommendations of airport management. Land use authorities suffer few, if any, consequences when they approve noise-sensitive land uses in high aircraft noise areas. Potential liability for Fifth Amendment taking claims appear to be more prominent in their thinking. The Fifth Amendment of the constitution states, in pertinent part:

> … No person … shall be deprived of life, liberty, or property, without due process of law; nor shall private property be taken for public use, without just compensation.

Claims and associated liabilities for takings fall upon the land use authority, not the airport. Note also that land use authority is given local planning authority through state-enabling legislation, and that these laws vary substantially from state to state. What can be accomplished by zoning boards in California may not be possible in Texas. For example, California requires a general plan that includes a zoning element and noise element for every municipality. Zoning at the county level is county-optional in Texas. The following paragraphs address the general mechanisms for implementation of compatible land use planning when a land use authority does implement land use controls for aviation noise.

Land use and development controls which are based on a well-defined comprehensive plan are among the most powerful tools available to local government units managing land use compatibility. These controls are beyond the authority of airport managements to implement, so it is the responsibility of local government units with land use jurisdiction to implement them in order to protect airports from encroachment. Even if an airport is managed by the same unit of government that has land use control authority for land areas beyond the airport's boundary, little coordination and discussion may occur regarding land use controls that should be implemented, and land uses that are compatible with airport development.

The need for coordination of development plans among all parties involved cannot be over-emphasized. This is particularly important when more than one unit of government has land use control authority in areas beyond airport boundaries. Airports are in a particularly precarious position when they may be liable for noise intrusions but have no authority to control the types of land uses that are developed beyond their borders. Local government units must accept responsibility for ensuring land use compatibility in their planning and development actions. Even state governments have roles to play, particularly in providing enabling legislation to allow local government units to institute land use controls.

The most common forms of land use controls available to the local governments include: zoning, easements, transfer of development rights, building code modifications, capital improvement programs, sub-division regulations, and comprehensive planning. These are discussed in the following paragraphs.

5.4.1 *Zoning*

Zoning is the most common and traditional form of land use control used in the United States today. It controls the type and placement of different land uses within the designated areas. It is used to encourage land use compatibility while leaving property ownership in the hands of private individuals or business entities, thus leaving the land on the tax roles. Zoning is not applied retroactively and is not necessarily permanent. It is most effective in areas which are not presently developed, and in which compatible zoning can be encouraged. Zoning often appears to be a simpler matter than it actually is. A municipality can incur liability by "downzoning" when land is zoned not at its highest and best use, but to a lower valued use.

5.4.2 *Easements*

An easement is a right held by one party to make use of the property of others for a limited purpose. Two specific types of easements are usually referenced in airport planning: a positive easement which allows the generation of noise over the land, and a negative easement to prevent the creation of a hazard or obstacle on the property of another.

5.4.3 *Transfer of Development Rights*

The transfer of development rights involves separate ownership of the "bundle of rights" associated with property ownership. The concept involves the transfer of the right to develop a certain parcel of property of a certain density/intensity to another parcel of property under the same or separate ownership. This would allow the property that obtains the added development rights to develop to an intensity/density that is beyond that which would normally be allowed. The airport could also purchase these rights from the landowner and retain them or sell them to another landowner. This concept can be used to retain property in compatible uses and still compensate the landowner for his loss of development. The idea depends on market conditions of the area and (there is some disagreement on this point) upon the availability of state-enabling legislation authorizing the development of the concept at the local level.

Transfer of development rights can also be used within a planned community where noise contours may split a planned community. Underlying zoning may establish residential density limits. An alternative that meets both the developer's development goals and the airports need for a buffer zone is to transfer the residential units permitted by zoning but located in high noise zone to a parcel or parcels outside the high noise zone. The net number of residential units remains the same, but residences are moved from the high noise area to a now higher density nearby parcel outside the noise zone. An example of this is the planned community of Aliso Viejo in Southern California. The planned community was divided almost in half by the noise contours from Marine Corps Air Station El Toro. Based on the planning efforts by the land planners and the County, all residential units in the high noise zone were changed to compatible commercial, warehouse, and retail uses and the permitted density in the portions of the community not within the high noise zone were increased to maintain the same number of dwelling units. A diagram of this planning effort is shown in Fig. 5.3.

The Aliso Viejo planning effort is an interesting case study of aviation noise/land use compatibility planning:

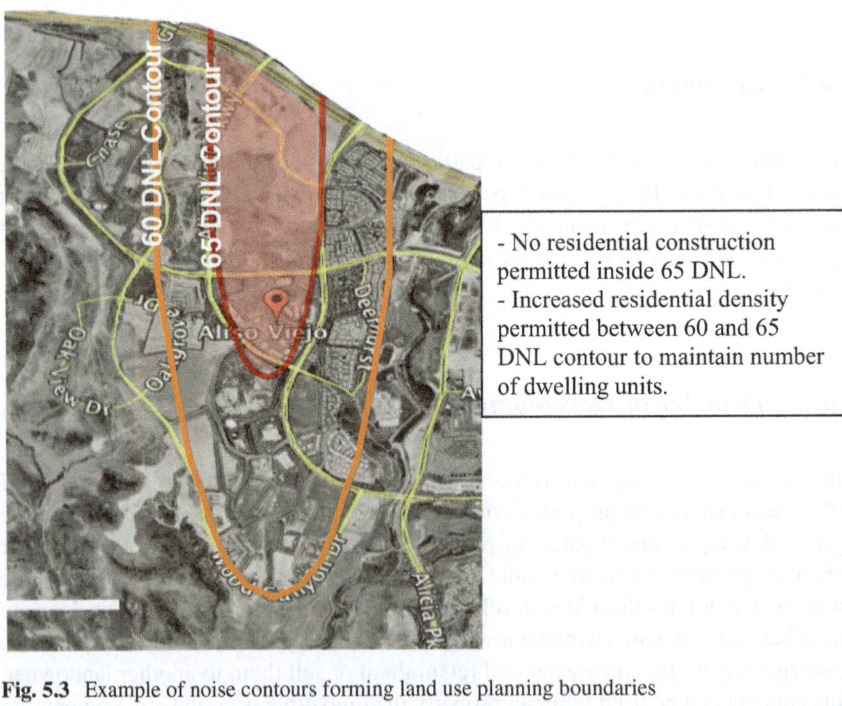

Fig. 5.3 Example of noise contours forming land use planning boundaries

1. The planning was done for a new community under the approach path to Marine Corps Air Station El Toro in the County of Orange where land value is very high. The military base was closed in 1999, but all of the planning and most of the land development occurred prior to the base closure. Further, the planning was done when the land was in an unincorporated area of Orange County. The land today is the City of Aliso Viejo.
2. The County wrestled for years with changing noise contours at El Toro as the base fluctuated in operational level based on world events. The problem was that long-term land use planning cannot be done when the noise contours change every year. Developers constantly came up with new noise studies that showed the noise contours were no longer on their property. The County resolved this issue by taking the noise contours in the military published noise study and declaring that the contour is a "policy line". This had the effect of eliminating annual fluctuations from the planning process as the line was mapped and adopted for use in all subsequent land use planning. This is an example of land use planning processes challenge from uncertainty in the location of noise contours and annual fluctuations in noise contours.
3. The 65 and 60 dB DNL[73] contours were used to form the land use plan. No residential use was permitted inside the 65 dB DNL contour. All of the residential units that could have been built were added to the allowed residential density outside the 65 dB DNL contour. The allowed much higher densities outside the 65 dB DNL contour than would have otherwise been permitted. Thus, the developer was allowed to build the same number of units. The center core, within the 65 dB DNL contour, became the "Town Center" with commercial, retail, and other non-residential uses.
4. The 60 dB DNL contour formed another boundary and the boundary road at the southern end of the community follows the contour. High-density residential was permitted between the 65 and 60 dB DNL contour. These apartments, condominiums, and zero lot line homes would be expected to have smaller outdoor spaces, thereby reducing exterior exposure to high aircraft noise levels.
5. All homes built in the 60–65 dB DNL range had to have architectural noise studies done to prove the design would meet interior DNL values of 45 dB, and required ventilation systems to allow for adequate ventilation with windows closed.

5.4.4 Building Code Modifications

This alternative modifies building codes to include specific sound attenuation provisions for structures within areas impacted by aircraft noise. This measure is effective for indoor noise level mitigation, but does not address exterior noise levels.

5.4.5 *Capital Improvements Programs*

A capital improvement program is a document that establishes priorities and costs for the funding and development of public facilities. It can be used successfully, in concert with sub-division regulations and a comprehensive plan, to control not only the areas of development but the timing of development by controlling the timing and location of public facilities. This is used for schools, libraries, or other noise-sensitive public facilities.

5.4.6 *Sub-division Regulations*

Sub-division regulations are used to control the design and placement of public and private facilities in the conversion of raw land to developed property. This is a key part of noise land use compatibility planning as these regulations are used to locate new communities and require that locations consider the context of aviation noise.

5.5 Sound Insulation Programs for Existing Homes and School

Generally, as part of the easement acquisition process, airport proprietors may institute a program to install sound insulation in homes subject to specific noise levels. The requirements for qualifying for and implementing a sound insulation program are contained in Appendix R "Noise Compatibility Planning/Projects" of the Federal Aviation Administration Order 5100.38D Airport Improvement Program Handbook (AIP Handbook). Specifically, structures must have an existing exterior noise exposure >65 dB DNL and an existing interior noise exposure >45 dB DNL in order to be eligible for a sound insulation program funded under Airport Improvement Program. This is in addition to having a current approved FAR Part 150 program. Sound insulation is a very effective program for reducing indoor sound levels, as sound rated windows and doors provide a significant reduction in outdoor-to-indoor noise levels. However, sound insulation provides no benefit for outdoor areas.

Many airports in the United States have implemented a residential sound insulation program (RSIP) to mitigate the impacts of aircraft noise. Unfortunately, no database documents the number of homes that have been part of an RSIP in the last 50 years. The FAA does publish the Airport Improvement Program funds expended each year on these programs, but these statistics do not go back to the original programs in the 1960s and 1970s, nor does this indicate the number of homes. For example, in the fiscal year 2016, FAA awarded $111 million to 12 airports for RSIPs. Of this total, $20 million was awarded to the City of Chicago for an RSIP around ORD (ANR Nov. 2018). The following are some examples of other large airport RSIPs.

5.5.1 Los Angeles

The City of Los Angeles began a residential sound insulation program in 1997 that has insulated more than 7,300 homes at the time of this writing. RSIPs have also been undertaken in cities adjacent to LAX outside the city of Los Angeles, including El Segundo, Inglewood, and portions of the unincorporated County of LA (https://www.lawa.org/en/lawa-environment/noise-management/sound-insulation-grant-program).

5.5.2 San Diego

San Diego began an RSIP in 1998 that has insulated 3,453 homes (http://www.san.org/Airport-Noise/Quieter-Home-Program) at the time of this writing.

5.5.3 Seattle

The Port of Seattle began an RSIP in 1985 which insulated more than 9,400 homes (https://www.portseattle.org/Environmental/Noise/SoundInsulation/Pages/default.aspx).

5.5.4 San Francisco

San Francisco RSIP program started in 1983. More than 15,000 homes have been completed to date (https://www.flysfo.com/community/noise-abatement/residential-sound-insulation-program).

5.6 Purchase Assurance Programs

Purchase guarantees can be applied to residential properties to help assure their salability. Sales agreements should assure that all future purchasers are cognizant of the noise levels and sign appropriate releases or easements. The advantages of this strategy are its retention of otherwise viable residential areas, allows residents the freedom to choose to stay or relocate, and maintains the stability of the neighborhood that might otherwise be blighted by vacant homes acquired through condemnation or sale to the airport proprietor. The process can be complicated requiring multiple appraisals and disputes that may occur regarding purchase price, that is, the value of the property in the absence of airport noise.

5.6.1 Indianapolis

Indianapolis Airport Authority implemented a purchase assurance program which involved buying, insulating, and reselling 103 homes (Indianapolis Business Journal, 2006).

5.6.2 Newport Beach

At Orange County Airport, a total of 464 residences in the Santa Ana Heights area have been purchased or otherwise made compatible through the County's Purchase Assurance and Acoustical Insulation Programs, circa 1985 (County of Orange 2017).

5.7 Acquisition of Land or Interest Therein (Easements)

The most community-invasive method to control and mitigate noise intrusion is to purchase noise-impacted property. Large-scale purchase programs can disrupt existing communities by inducing residential and commercial neighborhood blight. Blight can be a particular problem when some residents agree to a buyout, but others do not; when the areal extent of a purchase program strictly follows noise contours rather than natural community boundaries; and when local jurisdiction tax bases are compromised. Property purchase programs are also the most costly mitigation measure.

Acquisition by airports of some residences in the neighborhood but not others is difficult to implement. One method of keeping the area on the tax rolls and stable is to purchase the property and then resell it for a compatible use or to resell it for residential use while retaining a portion of the "bundle of rights" that are part of property ownership (see Purchase Assurance above). In other words, airports can re-sell purchased property to original homeowners or anyone else, but retain a covenant or easement preserving the airport's right to create overflight noise on the property. Such easements require property owners to forfeit their rights to initiate litigation against the airport for noise intrusion. In addition, this method allows the market to set the price and value of the noise easement retained by the airport. No matter what interest in land or property is purchased, the provisions of the Uniform Relocation Assistance and Real Property Acquisition Policy Act of 1970 (URARPAPA, PL 91-646) must be followed if federal assistance is involved in the purchase.

Land acquisition was far more popular in the 1960s and 1970s when homes were exposed to very high aircraft noise levels ($> L_{dn} = 70$ dB). Whole sub-divisions were acquired, for example to the north and west of LAX, and to the south of SEA. Land acquisition is much less common today, unless the land is being acquired for airport facilities.

The notion of limiting land uses to those uses that do not pose risks to the continued operation and expansion of airport in order to protect the public's investment in airport infrastructure is fraught with difficulties and limitations. Protecting the airport for future expansion by limiting development in one area allows for the impact of communities not protected by land use controls, that is, those communities built prior to implementation of land use controls. It may not be in the interest of these potentially impacted communities to support land use controls around the airport. Historically, land use controls have enjoyed limited success, particularly in the vicinity of major urban airports where development pressure is high, land value is high, and development approvals are subject to the vagaries of local politics. Most major urban airports were already surrounded by residential land uses at the time of the introduction of jet air transport creating what is an intractable planning problem (SEA, ORD, LAX, JFK, LGA, SFO, etc.).

Chapter 6
Airport Noise Mitigation

In the context of aviation noise control, mitigation is the practice of providing noise relief to communities already affected by aviation noise. The general practice has been to focus on acoustic measures to reduce the noise impact, even though both acoustic and non-acoustic factors contribute to adverse response to noise. Since very little effort has been made to manage non-acoustic determinants of community response to noise, this section addresses only acoustic factors.

6.1 Noise Control at the Source

Noise control at the source has been responsible for the greatest reductions in aircraft noise exposure since the introduction into service of jet air transports. These include noise reductions due to operational controls such as preferential runway use schemes, noise abatement takeoff and approach procedures, and modified flight tracks. The number of residences nationwide with aircraft noise exposures greater than $L_{dn} =$ 65 dB is one measure of the extent to which aircraft noise has been reduced by advances in engine design and aircraft aerodynamics from the first generation of jet transports. Figure 6.1 shows that from 1975 until today, the airport-vicinity residential population living within the current definition of aircraft noise compatibility ($L_{dn} =$ 65 dB) fell from about 7 million to just over 300,000 (the proportional change in exposed population is likely similar for $\mathrm{L_{dn}} = 55$ dB and $\mathrm{L_{dn}} = 50$ dB values. The absolute numbers are obviously greater at lower levels). Over the same time period, the number of passenger enplanements nearly doubled. Note that the change in exposed population size may be misleading, because it is critically dependent on the definition of "significant" noise impact.

Most of the reduction of aircraft source noise has come from advances in engine design, and from retirement of earlier generation aircraft. The bulk of the reduction in source levels was achieved by the year 2000, when FAR Part 36 Stage II aircraft (notably, the Boeing 727) ceased flying in the United States. The JT-3D engines of the

S. Fidell and V. Mestre, *A Guide To U.S. Aircraft Noise Regulatory Policy*,
https://doi.org/10.1007/978-3-030-39908-5_6

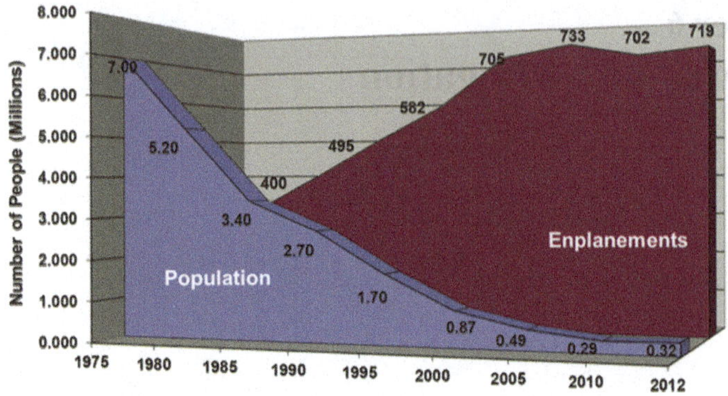

Fig. 6.1 Change in population size exposed to aircraft noise in excess of $L_{dn} = 65$ dB compared with change in aircraft operations. (FAA 2019)

first and second generations of transports (the Boeing 707 and 727) were low bypass turbofans with very high exhaust gas velocities and few nacelle treatments to reduce fan tones. Subsequent generations of high bypass ratio, low exhaust velocity turbofan engines for commercial transports emit far lower levels of broadband and tonal noise (Smith 1989). Certification noise levels as a function of the year of certification have decreased by 0.3 dB annually for decades, and continue into the present. (Annual operations, however, have increased on the order of 4–5%.)

Adoption of aircraft noise limits as part of ICAO Annex 16 Chap. 14, and Federal Aviation Regulation Part 36 has been responsible for much of the considerable progress in quiet engine technology. The rate of adoption of new technology has also been greatly influenced, however, by a concomitant improvement in fuel economy. International regulation of aircraft noise limits is governed by the treaty which established ICAO in an effort to coordinate aircraft design standards under the auspices of the United Nations. Member nations adopt regulations consistent with ICAO standards to achieve this harmonization. In the United States, FAR 36 establishes noise limits that aircraft manufacturers must meet to sell aircraft. Other member nations adopt the same noise limits specified in ICAO Annex 16, Chap. 14 to the convention on civil aviation so that aircraft can be sold internationally.

6.2 Noise Control Through Operational Restrictions

The sub-sections below describe restrictions that the public generally believes are available to airport proprietors to mitigate aircraft noise. Such restrictions were arguably available to proprietors prior to passage of the Airport Noise and Capacity Act of 1990, but have since been effectively prohibited. Most of these measures would today require the proprietor to comply with FAR Part 161, which implemented those parts of ANCA that addressed operational restrictions. The requirements of FAR

Part 161 are so onerous for airport proprietors that operational restrictions are for all practical purposes no longer available to airports. One of the goals of ANCA was to minimize the burden of higher costs and inefficient use of aircraft on the nation's air transportation system from a lack of consistency among individual airport mitigation measures. Thus, the operational restrictions noted in the following sub-sections are only notional, and can only be implemented by means of special legislation or by FAA itself. The public often reads the requirements of Part 161 as providing a path to implementing these procedures. This is a very misleading characteristic of ANCA and Part 161. The requirements of Part 161 form a kind of "Catch 22" where there is no practical way to meet the requirements of Part 161 and implement a noise-based airport access restriction. ANCA grandfathered rules in effect prior to 1990 and so a handful of airports have been able to implement some of these measures.

6.2.1 Denial of Airport Use to Aircraft not Meeting FAR Part 36 Requirements

FAR Part 95 phased out Stage 2 aircraft from the US fleet as of 1 January 2000. A similar phaseout off Stages 3 and 4 aircraft may eventually be considered as well. However, most Stage 4 aircraft meet Stage 5 noise limits, so a Stage 4 phaseout would accomplish little.

6.2.2 Airport Capacity Limits Based on Noise ("Noise Budgets")

These measures have been severely limited by the 1990 ANCA. The following paragraphs describe three types of capacity limits based on noise.

Restrictions based on cumulative noise impact

Long Beach Airport in California is the only commercial air carrier airport in the United States to have implemented a mandatory ("noise budget") restriction of this type. Long Beach's restriction is grandfathered under ANCA, since it adopted the measure prior to the passage of ANCA, following litigation ending in 1990. Long Beach Airport sets a minimum number of daily departures, and allows more departures as noise levels decrease. The cumulative noise level used in the Long Beach budget is CNEL, the California variant of DNL, as measured at the closest residential areas on each end of the airport's main runway.

Restrictions based on certificated single-event noise levels

This measure is severely constrained as an option due to the limits of ANCA and FAR Part 161. Certificated single-event noise levels by aircraft type are published by the FAA (AC-36-1H and AC-36-3H). Under such a restriction, an aircraft type can operate at an airport only if it has a certificated single-event noise limit that is lower than a limit set by the airport. Jackson Hole Airport, an airport located in Grand Teton National Park, is an airport with such an operational restriction.

Restrictions based on measured single-event noise levels

This measure is severely constrained as an option by ANCA and Part 161. Four airports in the United States have this type of operational restriction, all of which adopted the restrictions prior to passage of ANCA. These are Santa Monica Airport, John Wayne Airport, Torrance Airport, and Long Beach Airport, all in the State of California. These airports adopted noise limits enforced at permanent noise monitoring locations located around the airport. Both daytime and (stricter) nighttime limits are enforced. These programs are quite successful, having eliminated the noisiest aircraft from operating at the airport and preventing noisier ones from serving the airport, particularly at night. These programs are labor-intensive requiring a noise monitoring system and staff that verifies the identification of each aircraft associated with each aircraft noise event through the review of air traffic radio transmissions and radar flight tracks.

6.2.3 Landing Fees Based on Noise

More than 100 European airports have adopted noise-based landing fees since Zurich introduced the first in 1980 (EASA, https://www.easa.europa.eu/eaer/topics/market-based-measures/airport-charging-schemes). In the United States, however, ANCA and Part 161 prohibit noise-based landing fees. A noise-based landing fee can either accelerate the transition to quieter aircraft, or discourage use of noisier aircraft at an airport. Noise-based landing fees can fund other noise mitigation programs. While this program has been discussed often, it has never been implemented in the United States.

6.2.4 Complete or Partial Curfews

Implementation of night curfews as a new restriction on flight activity in the United States is severely limited by the ANCA, and would require a Part 161 application for implementation. It is one of the most common requests from airport neighbors, and one of the most strongly opposed measures by the airline and cargo industry. Airports with curfews (grandfathered under ANCA) in the United States include

Santa Monica, John Wayne, San Diego, Long Beach, and Jackson Hole. Frankfurt airport in Germany recently adopted one of the strictest curfews in the world. Frankfurt's curfew varies the permitted hours of aircraft operations on the basis of noise level and prohibits all landings between 2300 and 500 h. One side effect of the Frankfurt curfew was an increase in night operations at other airports in southern Germany and the Netherlands.

6.2.5 *Noise Barriers (Shielding)*

Noise barriers are not practical and are only of value for engine runup areas. These could include staging areas prior to entrance onto the runway before takeoff or maintenance runups. Once an aircraft is airborne a barrier is ineffective.

6.2.6 *Noise Control Through Aircraft Operational Procedures*

The sub-sections below describe operational procedures that people generally believe are available to airport proprietors to mitigate aircraft noise. In reality, these measures are the exclusive authority of the FAA, and entail either air traffic or aircraft operational procedures (the responsibility of FAA's Air Traffic Organization and FAA Flight Standards Services, respectively). Consideration of operational mitigation measures is part of the Noise Control Program of the airport FAR Part 150 program.

6.2.7 *Departure Thrust Cutback*

This measure cannot be implemented without the direct concurrence of the FAA and compliance with Advisory Circular 91-53A. Known as Noise Abatement Departure Procedures (NADP), thrust cutbacks can be used to reduce noise to certain populations. Describing them as noise abatement can be misleading as they do not reduce total noise, but relocate its geographic distribution from one area to another. FAA has defined two NADPs: one for noise reduction close-into the airport (close-in NADP) and one for noise reduction more distant from the airport (distant NADP).

The concept for a close-in NADP is that an early thrust cutback will reduce noise near the airport, because lower thrust generates lower noise levels. The lower thrust also results in reduced climb gradient, so that as the aircraft gets farther from the airport the aircraft is lower and noisier than the no cutback case. AC 91-53A requires that the close-in cutback occurs no lower than 800 feet above field elevation, and that the resulting cutback thrust would result in a positive climb gradient (which depends on the number of engines and whether the aircraft is equipped with automatic thrust restoration) in the event of an engine shut down at the worst possible point in the takeoff.

The distant NADP uses normal departure thrust and the thrust cutback occurs after flap retraction farther down the flight track. This results in a steeper climb gradient, thus a higher altitude when the thrust is reduced and the community at that distance benefits from both higher aircraft altitude and lower thrust, both of which reduce noise levels. However, the reduction in distant noise levels is purchased at the price of higher noise exposure closer to the airport, due to maintenance of takeoff thrust for a longer distance in lieu of reducing power to climb thrust sooner. (Noise levels are very sensitive to the engine thrust levels.)

In ICAO terminology, the close-in and distant procedures are known as NADP1 and NADP2. In both the AC 91-53A and ICAO documents, the actual cutback thrust level is not specified, nor is the cutback altitude specified other than the minimum permitted altitude and thrust. Thus, neither NADP is standardized in the sense that any given airline/aircraft doing an NADP will produce a given noise reduction at a particular location.

AC 91-53A also mandates that an airline have only one close-in and one distant procedure per aircraft type for the entire fleet and for all airports, so NADPs cannot vary by airport (in terms of altitude of cutback, or cutback power setting). The most widely known and studied NADP is the close-in NADP used at John Wayne Airport. It is generally a cutback at 800 feet to a thrust that just maintains a positive climb gradient should there be an engine failure. The runway at John Wayne Airport is only 5,700 feet long, and the community that benefits is the community located just off the end of that runway. For most airports with longer runways, this community would be well within the airport boundary. Any airline using this NADP at John Wayne must use this NADP at any other airport in the United States if a close-in NADP is requested.

6.2.8 Flight Track Alterations

This alternative re-routes approaching and departing flight tracks to minimize noise exposure in sensitive areas. These re-routings are controlled by considerations of operational safety and air traffic control procedures. Generally speaking, such air traffic control procedures can be implemented, perhaps with penalties involving reductions in airport and airspace capacity. However, increases in fuel burn by aircraft making turns at low altitudes, where the aircraft are in a low-speed, high drag configuration, can adversely affect aircraft operating margins. Maneuvering during the last three to four miles of the final approach (in good weather), and within the final six to seven miles (during poor weather), is also undesirable, because pilots cannot establish and maintain a stabilized approach to landing.

As the FAA has implemented NextGen[74] procedures, flight routes are being modified across the United States. The drivers for new flight paths are efficiency of airspace use and reduced aircraft fuel burn. Since noise has not been a driving factor in NextGen/Metroplex projects, many regions affected by the NextGen upgrades have experienced severe adverse community response and litigation. These include communities in Phoenix, Southern and Northern California, Baltimore, and Washington D.C., among others. No policy has been developed for routing decisions to evaluate the tradeoff between efficiency of aircraft operations and noise impacts. Because the NEPA process essentially focuses all consideration of noise concerns to areas within the 65 dB DNL contours, those affected by NextGen are effectively left out of current environmental review requirements.

6.2.9 Preferential Runway Use

This alternative attempts to favor the use of runways that have less incompatible land use associated with its flight corridors. The FAA is responsible for implementing this program, based on the recommendation of the airport operator and the safety considerations contained in Federal Aviation Regulations Part 121. Winds generally dictate runway use, but in calm or near calm winds, particularly during slow traffic periods, a runway use program encourages the use of runways which avoid overflights of residential areas. Even airports with one runway can benefit from such a program if there is opportunity to avoid residences by reversing the traffic flow. This is a particularly useful program at night because of lower traffic volumes at night and generally calmer winds. At LAX, for example, approaches are conducted over the ocean at night, on runway pairs separated from departure runways by about a mile.

6.2.10 Power and Flap Settings

A variety of noise-reducing operational procedures can be implemented at airports. These include minimum flap landings and delays in flap and gear deployment. To help minimize fuel cost and flight time, most operators of large air carrier aircraft have adopted procedures for reduced flap setting and delaying flap and gear deployment, consistent with safety and current aircraft and air crew capabilities. During Visual Meteorological Conditions (VMC), weather conditions and low traffic conditions, air carrier aircraft generally lands with minimum flap settings at an airport. More extensive delayed flap procedures are not considered safe with current air traffic control procedures and safety criteria. Optimized performance descent (OPD) practices, formerly known as continuous descent approach, are of particular note.

A step-down approach, in which aircraft approaching airports fly long, level flight path segments between three-degree descent segments (sometimes called a stair step approach), is a common air traffic control procedure to separate aircraft. Higher thrust levels are needed to maintain a constant altitude on these level steps during approach, whereas quieter flight idle or near-idle thrust suffices for a continuous three-degree approach slope. Differences in noise levels on the ground beneath step-down and continuous descent flight paths can be substantial. Continuous descents are a key feature of the next-generation air traffic control system. Continuous descent flight paths are supported by the airlines (Clarke 2006) because they reduce fuel burn on approach.

Chapter 7
Potential Changes to Airport Noise Policy

Community noise issues remain the major impediment to airport capacity enhancements in the United States. This chapter focuses attention on policy issues of the greatest salience in contemporary airport/aircraft noise controversies. Various stakeholders have differing opinions about these matters, which policymakers have the duty to balance. Additional issues (including sustainability, climate change/flight shaming, and air quality) also affect the ability of airports to expand, and also play roles as non-acoustic variables affecting noise-induced annoyance.

The discussion of this chapter is based on the observation that airport noise management occurs at both federal and local levels. Arguably, the most important noise control policies are implemented at the local level, so several of the changes discussed in this chapter seek to return regulatory control, at least in part, to local authorities. This goal may require state-enabling legislation, support from modernized federal policies, and in some cases, relief from federal pre-emption of local programs.

The *status quo* of aircraft noise regulatory policy in the United States is technically obsolete and logically indefensible. Current policy is based on an accumulation of ad hoc decisions made from the 1950s onward, intended for circumstances of air commerce that have changed greatly over the decades. The bases on which these decisions have been made are long overdue for re-consideration. For example, promotion of civil aviation has ceased to be among the regulatory goals that Congress has directed FAA to pursue. The former Congressional direction was once the basis for decades of simplistic and lenient noise policies, justified by incomplete and (in retrospect) demonstrably incorrect understandings of the effects of aircraft noise exposure on communities. These erroneous understandings are at odds with FAA's own vision statement,[75] not to mention inconsistent with current international technical consensus standards. Current policy does not provide equal protection from exposure to highly annoying aircraft noise to residents of all U.S. airport communities, and erroneously supports oft-repeated claims of diminishing long-term aircraft noise impacts in airport neighborhoods nationwide.

S. Fidell and V. Mestre, *A Guide To U.S. Aircraft Noise Regulatory Policy*,
https://doi.org/10.1007/978-3-030-39908-5_7

More to the point, however, U.S. noise regulatory policies rarely achieve their own goals. (Successes in reducing the population within the $L_{dn} = 65$ dB exposure level have been major, but are attributable in large part to a combination of internationally harmonized regulation and pursuit of lower fuel burn aircraft engines.) Community opposition to airport operation and growth, exacerbated by inefficient regulatory protection of airport neighborhoods from exposure to highly annoying aircraft noise, has delayed and complicated construction of new infrastructure at existing airports (Fidell and Mestre 2019). No new major airport in the United States has opened on a greenfield site since 1995. This chapter discusses a number of potential modifications of U.S. aircraft noise policies that could help to further development of a safe and efficient national air transportation system by easing aircraft noise controversies and community opposition to greater airport capacity.

7.1 Basis for Current Aircraft Noise Management Policies

A 2015 letter to FAA from eight aviation industry trade associations congratulates FAA for its "fact- and science-based" approach to regulation of aviation noise: "We [*Airlines for America, Air Line Pilots Association, Aerospace Industries Association, Cargo Airline Association, General Aviation Manufacturers Association, National Business Aviation Association, National Air Carrier Association, and the Regional Airline Association*] write to express our appreciation for the fact- and science-based approach FAA takes in addressing aircraft noise…".[76] As described below, this position is untenable, even though FAA has set a goal of reviewing current policies on the basis of the best available science: "Noise impact and mitigation criteria and land use compatibility guidelines must be based on the best available science" (Sizov and Cointin 2011).

There is less than meets the eye to assertions that current U.S. aircraft noise regulations are justified on technical grounds. In fact, the claim fails to recognize von Gierke's (1973) observation that regulatory policy is not in fact based primarily on technical evidence: "…the main and basic problem … is a social, ethical, and economic one: what percentage of the population shall be protected and at what price". However much it may be asserted that aviation noise regulatory policies are "fact- and science-based", the reality is that regulatory policy is based in large part on tradeoffs among stakeholders which are properly informed by the latest scientific understandings. Given both the changes in FAA's congressional direction, and advances in scientific understanding over the last quarter century, modernization and improvements of aircraft noise regulatory policy are long overdue.

Policy positions inevitably involve non-technical value judgments about economic and political matters. Consideration of complex policy issues should be informed and defended by the best available technical information, not solely by case law. Whether for lack of technical understanding, or for reasons of judicial deference, courts seldom bother to closely consider technical aspects of regulatory policies. Because courts seldom consider technical analyses, case law defenses of regulatory policy reflect little more than generic judicial deference to federal agency positions.[77] Legal precedent is, however, a poor substitute for a substantive technical reasoning as a basis for regulatory policy. In effect, case law defenses of policy positions amount to little more than the observation that "Some courts have accepted similar arguments in the past", rather than "Logic and evidence strongly support FAA interpretations of technical findings". The net result is that technically indefensible policies persist simply because some courts have tolerated them, rather than because of the aptness and technical correctness of the policies themselves.

While scientific understanding can lead to improved predictions of the effects of aviation noise, no step-function measure of noise impacts can ever support a bright line at a single noise dose that distinguishes "acceptable" from "unacceptable" levels of aircraft noise exposure. At best, judgments about the balance of conflicting interests may be made with awareness of relevant scientific information. Such judgments can hardly be said to be "based on" such information.

If an obsolete dose–response function predicts that 12.3% of the residential population is highly annoyed by transportation noise at an exposure level of L_{dn} = 65 dB, how does it follow "scientifically" that a DNL value of 65 dB can serve as a policy threshold separating insignificant from significant levels of aircraft noise exposure? If a modern dose–response function endorsed by an international technical consensus standard (ISO 1996-1, 2016) indicates that 12.29% of the population is highly annoyed specifically by aircraft noise at a DNL value about 8 dB lower than 65 dB, does it follow "scientifically" that FAA's policy regarding the threshold of significance must, of necessity, be reduced by 8 dB? This could only be the case if a prior decision had first been made to accept 12.29% of the population as highly annoyed as appropriate, and then determine the noise exposure reflected by that level. In fact, the decision-making sequence for existing policy was just the opposite.

Consider, for example, the scant attention paid in the FICON (1992) report to evidence of the source-specificity of annoyance, and to non-acoustic influences on the annoyance of transportation noise. The technical literature contained many peer-reviewed reports of source-specificity of annoyance by the time that the FICON report was prepared (*cf*. Hall et al. 1979, 1980, 1988; Kryter 1984; Job 1988), as well as indications that non-acoustic factors affected annoyance judgments (*cf*. Borsky 1983; Fidell et al. 1988). The implications of these findings were largely ignored in favor of a developing a dose–response relationship by means of selective, univariate logistic regression, using noise exposure level alone as a predictor variable, for the annoyance of transportation noise of all sources.

Interest in reviewing and revising the national noise policy was renewed in 2010, when FAA began a "roadmapping" process to update its aging noise policies.[78] This process has proved to be a slow and often delayed one, and is far from over.

7.2 Measures Capable of Improving Airport/Community Relations

One institutional circumstance that can foster poor airport/community relations is political insulation of the airport proprietor from the population affected by noise created by its airport operations. Only a few communities exposed to the aircraft noise created by ORD vote for the mayor of Chicago; San Francisco's airport is located neighboring San Mateo County; Cincinnati's airport is located not only in another state, but in a different FAA region; and so forth. In principle, airport proprietors and aircraft noise-exposed nearby residents should be contained within the same political jurisdiction to permit meaningful local control of airport noise.[79] Although politically difficult to achieve in some cases (since it can require a change in airport management), exposing airport proprietors to noise-exposed electorates could relieve some of the bitterness that can result from feelings of powerlessness of noise-exposed populations.

FAA's charter might need to be revised to return a limited range of authority and a modicum of control of airport development plans to municipalities and airport proprietors (as specified by the Air Commerce Act of 1924). Nonetheless, simply giving local jurisdictions other than that of the airport proprietor a meaningful voice in plans for airport development would encourage the sort of compromise and flexibility that could ease airport noise controversies.

Freestone and Baker (2018) suggest creation of some form of multi-level, inter-governmental agency focused on airport/community relations. The unaddressed devil, as always, is in the details. Similar approaches, such as airport roundtable groups, have failed to substantively ameliorate airport/community controversies in the United States.[80]

Some states permit at least some regional oversight of development plans near airports. California requires a higher body review of development within the airport environs (Public Utilities Code, Section 21001). This higher body is a County-level Airport Land Use Commission. The commission establishes a land use plan for the airport environs. Municipalities submit development plans to airport land use commissions for determinations of consistency. If found inconsistent, a municipality may over rule the commission with a 4/5 majority vote.

Texas has enacted an Airport Zoning Act (Texas Local Government Code, Section 241.001). This act does not rely on noise contours to define the review area for local development projects, but defines a box that extends 5 miles from the end of each runway and 1.5 miles on either side of runway centerlines. The act defines a review process for development within this area.

Hawaii has a statewide land use review process. Zoning on the local level is done by the counties (each of the Hawaiian islands is a county). In addition to the local zoning, zoning plans are reviewed by the State Land Use Commission. The State Department of Transportation, Airports Division, has adopted noise/land use compatibility regulations that are stricter than the FAR Part 150 guidelines (60 DNL instead of 65 DNL for residential compatibility). However, for airport Part 150 programs within the State of Hawaii, FAA only permits the plan to address the federal guideline of 65 DNL, and relegates any discussion of the State's 60 DNL guideline to an appendix.

Some other measures whose adoption might help to improve the logic and effectiveness of aircraft noise regulation in the United States are noted in the following sub-sections. The measures discussed are generic, and obviously require refinement and elaboration before implementation.

7.2.1 Adopt a Systematic Rationale for Defining the Significance of Noise Exposure

As described by Fidell (2015), FAA's technical rationale for identifying $L_{dn} = 65$ dB as a threshold of significance of noise exposure has never been rigorously laid out, nor defended on the basis of modern technical understandings of community response to aircraft noise. This threshold value, converted from 1950s-era CNR units first into units of NEF in the late 1960s, and subsequently into units of DNL, can be traced to opinions offered by a few prominent acoustical consultants in the early 1950s. The goal at the time was not to protect communities from exposure to highly annoying noise, but rather to limit cumulative aircraft noise exposure to levels that could keep complaints about aircraft noise on military airbases to manageable proportions.

Post-ASNA and FAR Part 150, the goal of regulation has morphed from minimizing complaints about military aircraft noise to protecting civil residential populations from exposure to highly annoying aircraft noise. No official or definitive analysis has been offered, however, of why the same threshold value that was appropriate for minimizing complaints about military aircraft noise in the Cold War era is coincidentally appropriate for protecting some specifiable percentage of today's residential populations from exposure to highly annoying aircraft noise.

An explicit rationale for expressing a criterion of the significance of aircraft noise exposure in decibel units is also lacking.[81] If the goal of modern regulation is truly protection of residential populations from exposure to highly annoying aircraft noise exposure, then it is far more straightforward to define significance directly in terms of the percentage of the population protected from high annoyance. A case, however, needs to be made, whether by formal cost/benefit analysis, or by other quantitative means, for whatever value FAA believes is appropriate. Lacking such justification, defining $L_{dn} = 65$ dB as a threshold of significance of noise exposure is essentially arbitrary.

Clearly, protecting very few and protecting everyone from exposure to highly annoying aircraft noise are extreme and unreasonable goals. A formal argument has never been constructed for considering a level of noise exposure that annoys 12.29% of the population—the percentage of the population of a hypothetically average community that FICON (1992) believed was associated with a DNL value of 65 dB—as a threshold of significance. Consider a map with contours labeled with the associated population highly annoyed. Reviewers, land use agencies, and potential buyers would see contours with labels of, say, 75, 50, 25, 10, and 5% highly annoyed. Such directly understandable information might more fully inform home purchase and zoning decisions than decibel-denominated contours. An airport and a land use agency could immediately and directly appreciate the size of the population that would be affected adversely, and act accordingly.

It is now internationally accepted (per ISO 1996–1, 2016) that the actual prevalence of high annoyance with aircraft noise exposure in a hypothetically average community is more than twice as great as thought by FICON a quarter of a century ago. If the goal of regulation is protection of some percentage of the population from exposure to highly annoying aircraft noise, the threshold of significance of noise exposure, as expressed in units of decibels, should be revised to conform with the goal.

7.2.2 Abandon One-Size-Fits-All Dose–Response Policy

Despite the well-documented great variability of annoyance prevalence rates among communities with the same aircraft noise exposure (see Figs. 3.5 and 3.6), FAA applies a single, fixed definition of the "significance" of noise exposure to all. This practice guarantees that different communities will experience different levels of exposure to highly annoying aircraft noise. As Fig. 3.7 shows, limiting aircraft noise exposure to $L_{dn} = 60$ dB would protect virtually every resident of some communities from exposure to highly annoying aircraft noise, but protects only about 10% of the residents of other communities from such exposure. A fixed, single-valued, nationwide definition of the significance of noise exposure thus deprives people in many communities of equal effective protection of the law.

Three common counter-arguments to community-specific definitions of the significance of aircraft noise exposure are:

First, differing definitions of the significance of aircraft noise exposure in different communities would conflict with providing uniform protection to all communities;
Second, it could lead to the dreaded (by airlines) "patchwork quilt" of regulation; and
Third, it would encourage efforts by community activists to exaggerate the local prevalence of high annoyance with aircraft noise exposure.

The first argument is simply specious. It is the current, fixed, nationwide policy—not a community-specific policy—that fails to provide residents of different communities with equal protection from exposure to highly annoying aircraft noise. The issue of uniformity of application of regulation is properly one of uniformity of effect, not uniformity of a particular numeric value of noise exposure. As described by Fidell et al. (2014), a fixed $L_{dn} = 65$ dB definition of the significance of aircraft noise exposure is wasteful and inefficient. Regulation of aircraft noise exposure in communities that are willing to tolerate greater exposure to aircraft noise is unnecessarily stringent. On the other hand, communities that are less tolerant than average of aircraft noise are not afforded sufficient protection to lessen opposition to airport operation and growth.

None of this implies that internationally harmonized aircraft noise certification levels should be anything other than uniform across the United States. But in matters of defining locally acceptable noise limits for land use control, a suggestion permitted in a footnote (quoted in Sect. 5.3) to the current Part 150 land use/noise compatibility table, there is little to justify a national norm in light of knowledge that communities form unique attitudes about noise.

The second ("patchwork quilt") argument is less cogent today than it was decades ago, when airline fleets were composed of more noise-heterogeneous aircraft types. The argument made more sense when the fleets serving individual airports still contained large proportions of Stage I and II aircraft, and before curfews, maximum levels, and before other noise mitigation measures were effectively prohibited by ANCA. When the fleet was composed of aircraft producing very different noise emission levels, the constraints on route planning could be pronounced. Today, the bulk of the large commercial passenger jet fleet is Stage 3 and Stage 4-compliant, and differs little by aircraft type. Some measures that could contribute to this "patchwork quilt" argument are limited to real-time noise level limits at a few airports (e.g., SNA and LGB) or to restrictions on aircraft stage lengths. If ANCA permitted, United States airports, like European airports, airports and communities might be better served by basing landing fees on noise rather than on noise performance restrictions, and letting airlines evaluate the cost-effectiveness of operations and fleet decisions.

Perhaps recognizing the differing roles of federal, state, and local governments would enhance the discussion of national noise policy. Such a recognition would include a re-evaluation of state and local roles in aircraft noise management, consistent with a nationally safe and effective air transportation system.

7.2.3 Repeal or Revise ANCA

The 1990 U.S. Airport Noise and Capacity Act (ANCA) effectively prohibits airports from restricting aircraft operations in any meaningful way. By depriving airports of flexibility in reaching compromises with surrounding communities about their aircraft noise exposure, ANCA has hardened community opposition to airport expansion nationwide, while raising the degree of controversy of proposals for constructing airport infrastructure (Fidell and Mestre 2019). ANCA has provided virtually no noise relief to communities that would not have occurred long ago by means of the economically-motivated retirement of Stage 2 aircraft. At best, ANCA has preserved existing airport capacity, while providing little new capacity.

Although noise-based opposition to airport expansion antedates ANCA, the net effect of ANCA's prohibition on operational restrictions is that it has become nearly impossible to increase airport capacity in the United States, due to nationwide community frustration with the process. It is important to note that all U.S. airport development is initiated by, and the responsibility of, individual airport proprietors. Without the sort of initiative from the local airport proprietor that ANCA discourages, new runway construction projects are unlikely.

7.2.4 De-Linkage of Policy Thresholds

Aircraft noise exposure at levels in excess of $L_{dn} = 65$ dB currently serves not only as a threshold of significant noise exposure, but also as an eligibility requirement for participation in residential acoustic insulation programs, and as a land use compatibility recommendation. The same threshold level of aircraft noise exposure does not necessarily have to serve all of these regulatory purposes. For example, sound insulation programs for homes and schools are intended to yield an interior environment suitable for speech communication. Linking eligibility to the outdoor noise level is unnecessary, and relies on antiquated assumptions about the outdoor to indoor noise reduction of typical structures. Tying the eligibility of sound insulation to outdoor noise levels unnecessarily links sound insulation costs to a lowered outdoor noise policy. The linkage between outdoor and indoor noise levels should be discontinued. A home, school, theatrical, religious, or other noise-sensitive indoor use should be eligible for sound insulation based solely on interior aircraft noise levels.

The knee jerk reaction to discussions of reducing the national noise guideline of $L_{dn} = 65$ dB to a lower level has been negative, due in part to an assumption that the indoor noise guideline would be similarly reduced, which would greatly expand eligibility for residential sound insulation. The residential sound insulation link to the outdoor noise exposure is tenuous at best, relying on broad assumptions. Indoor noise guideline and sound insulation eligibility are a separate and distinct topic. Consider, for example, the amount of residential development that has occurred within the 60–65 DNL contours since 1975, arguably the date at which land use planning should have been better informed. Opposition to the $L_{dn} = 60$ dB guideline has been based in part on aversion to greater costs of residential sound insulation programs.

7.2.5 Adoption of a Regional Focus on Airport Capacity

Airport planning is routinely conducted on an individual, airport-by-airport basis. A regional, rather than individual airport, perspective would help to improve the efficiency of the national air transportation system, while avoiding wasteful overbuilding of capacity at any single airport. Since airports compete with one another for air traffic, the regionally uncoordinated nature of such planning can lead excess capacity in some areas, and under-capacity in others. It can also lead to boom-or-bust, winner-take-all airport economics as airlines change their hubbing operations, as for example, when Cincinnati, Dayton, Detroit, Memphis, and Minneapolis were all competing for Delta traffic.

Airlines prefer to concentrate their operations (and hence, the noise exposure that they produce) at a small number of large hub airports. Individual airports sometimes prefer to provide specialized services (e.g., by building air cargo support facilities and ramp parking areas to attract night cargo operations), even when their nighttime runway capacity might be more advantageously used regionally to support, say, additional long-haul passenger flights.

When airports assess the environmental impacts of their proposed infrastructure projects to satisfy NEPA disclosure mandates, NEPA requires that they consider not only the impacts of the preferred alternative, but also the "no action" and several plausible alternatives. FAA could further require that, whenever feasible, one of these plausible alternatives should be construction of additional capacity at a nearby airport rather than at the project sponsor's airport. Because project sponsors at one airport usually lack the authority to engage in development at nearby airports, additional changes would probably also be required to project review procedures. These could include a requirement for FAA to consider the regional purpose and need of individual airports' development plans before awarding funding.

By replacing INM with AEDT modeling of impacts of aircraft operations, FAA has already taken a step helpful for assessing the regional needs for airport planning, multi-airport systems, and even the global impact of aviation. (INM is predicated on predicting noise exposure produced by a single airport's operations. AEDT permits much broader consideration of the combination of multiple airports' operations.) FAA's responsibilities for providing a safe and efficient *national* air transportation system are better served by a system in which airport improvements are *not* proposed, evaluated, and approved on an individual airport basis.

7.2.6 Revision of Revenue Diversion Restrictions

FAA interprets its Congressional direction (per the Airport and Airway Improvement Act of 1982, the Airport and Airway Safety and Capacity Expansion Act of 1987, the FAA Authorization Act of 1994, and the FAA Reauthorization Act of 1996) as forbidding uses of public airports' revenue for purposes other than airport capital or operating costs, as well as the costs of other facilities owned or operated by the sponsor and directly and substantially related to air transportation (FAA 2009). FAA also forbids airports which have received federal assistance from diverting such revenue, or even from closing, before the term of its grant assurances has expired.

The only airport revenue that is not subject to the revenue diversion restrictions is that produced by sales tax collected by the airport. Most sales tax is collected by concessionaires, however, and does not pass through airport hands. Similarly, property taxes for land or fixtures per individual state regulations are collected by the state, and do not pass through airport hands. Even when an airport elects to spend airport revenue on noise mitigation, federal law limits those programs to those areas within the $L_{dn} = 65$ dB contour.

Such policies bar airports from spending PFC revenue streams or other airport-derived revenue for noise mitigation purposes. Further, areas within which airports can spend revenue for noise mitigation are limited to those areas within the $L_{dn} = 65$ dB contour. The lack of funds for noise mitigation that the airport may use limits the ability of the airport to compensate residents for the disproportionate impacts on communities near the airport. Although the notion of compensation is avoided in the United States, it is probably the reason residential sound insulation programs are so popular with residents: funds to replace windows and doors and provide upgraded insulation and ventilation add value to homes. The forms of resource-heavy mitigation that an airport may use to build bridges to the community using airport revenue include the following:

- Residential Sound Insulation Programs: As mentioned above RSIP provide in effect a re-model of the homes sound insulation characteristics. Their costs vary widely over a range of tens of thousands of dollars depending on the age of the home (need to remediate lead paint and/or asbestos insulation), number of homes in the program (quantity discounts on window and door manufacturing are substantial), geographic location (temperate climate homes are generally less well insulated than homes in colder or hotter climates and more work is need to seal the home). In some programs the airport or cities may need to provide some assistance to homeowners whose home cannot be sound insulated until building code non-compliance on certain safety issues are corrected (most commonly unsafe electrical circuits, structural integrity due to rot, termites or shoddy construction).
- Residential Purchase or Purchase Assurance Programs: Wholesale acquisition of property, acquiring entire neighborhoods, due to noise impact is invasive and generally not a popular option. Voluntary programs have a risk of blight associated with vacant buildings and properties further impacting the community. Purchase assurance is a program where a homeowner claims that they want to move but the impact of noise has restricted their ability to sell at a fair price. Purchase assurance is a program where after multiple appraisals a fair price is derived and any difference between actual selling price and fair price is made up by the airport. The program would by necessity be complex and may not overcome other disincentives to sell. For example, in California, property tax growth is limited a set maximum yearly increase based on the purchase price or the cost basis of the home in 1976. A homeowner who has resided in an area for many years could move from a low tax rate to a very high tax rate by relocating. There are complex rules for transferring the basis to another home and the complexity may deter this alternative. The cost to implement a purchase program would be highly sensitive to geographic area. Average home prices across the United States range from under $200,000 to over one million dollars, so the economic feasibility of this program may vary across the United States This is an area where state legislation may be needed to effect these measures.

- Property Tax Rebate Program—Residents: Currently only one city in the United States offers such a program: the City of Rosemont, near Chicago O'Hare Airport. This type of program is difficult when property taxes are collected not by the airport, but by other municipal or state entities. Most such entities are not inclined to offer such a program that benefits the airport, so the mechanism to implement such a program would be difficult and controversial. Guidance for the airport on the level of rebate that would be reasonable for the amount of noise impacting the community is lacking. Such a program could raise the value of a home due to its lower property tax rate. This is another area where state legislation may be needed to effect these measures.
- Property Tax Rebate Program—Cities/Counties: In locations where land restrictions would result in lower tax and fee revenue, municipalities have a disincentive to regulate land use if the greatest revenue is derived from a non-compatible land use. In the land use decision process airports could provide compensation for lost revenue due to undeveloped or underdeveloped land near the airport. Since residential land uses are not usually the highest revenue generator due to the high cost of services to residential areas, the cost of this program may not be as onerous as it seems. On the other hand, with fewer municipal residents the revenue to other businesses in the community will be reduced.

Under current policy, programs such as the above can only be accomplished with FAA approval, and only within the $L_{dn} = 65$ dB noise exposure contour. Airports' problems with surrounding communities extend well beyond the $L_{dn} = 65$ dB contour.

7.2.7 *Closer Scrutiny of Airport Noise Exposure Contours Used for Land Use Planning*

AEDT is a deterministic aircraft noise modeling program whose predictions are fully controlled by the assumptions adopted for each run. The substantive issue in any prospective noise modeling exercise is thus not the location of predicted noise exposure contours, but rather the assumptions adopted to produce the contours. Since there are no facts about the future (it hasn't happened yet); since air traffic forecasting is an arcane and fundamentally uncertain enterprise; and since the best laid plans of mice and men are blind to unanticipated developments (changes in runway use patterns, airline schedules, fleet composition, departure and arrival routes, fuel prices, national economic cycles, etc.), communities have little alternative to accepting an airport's predictions of future noise exposure levels. This creates a difficult problem for land use planning and for land use authorities.

In the case where a municipality is preparing a land use plan for large undeveloped property near the airport, adoption of a noise contour as a residential boundary is straightforward. The land use plan can arrange residential, commercial, industrial, recreational, and education uses according to the overall master plan. Unfortunately, in 2019, at the time of this writing, there are few such large-scale master planning opportunities. Today, the evaluation of undeveloped land is limited to parcels undeveloped surrounded by other existing development. A not uncommon problem is infill residential. This is a case where an existing residential area located in a noise impacted area has available empty lots. It is not practical or desirable to limit such lots to non-residential uses, except perhaps for an occasional pocket park. These lots will end up in residential use, so at most an airport can seek easements, sound insulation, and disclosure notification.

Disputes over the location of a noise contour pose more difficult problems for planning jurisdictions. A sophisticated developer is likely to engage an expert who can dispute the methods and assumptions used to generate the contour, and point out the underlying uncertainty of the contour location. Since land use legislation varies idiosyncratically by state, a general solution may not be available. State-enabling legislation varies widely among the states from relatively strict land use regulations in California and the Northeast, to Texas where land planning is county-optional. Land use guidelines recommended by the FAA and the airports need to reflect local reality and uncertainty.

In one notable case described in Sect. 5.4.3, the Orange County, California General Plan, was written to describe an aircraft noise exposure contour simply as a "policy line". This terminology sidesteps disputes about where an airport's CNEL = 65 dB contour is actually located in any given year. The net effect is to adopt a residential development definition, rather than an explicit aircraft noise contour, for an area from which residential development is excluded (with the exception of infill development and caretaker's quarters). The approach has worked well in practice, because the policy line was adopted in 1985 when the aircraft fleet was largely noisy Stage 2 aircraft. Since Stage 2 aircraft were excluded from the fleet in 2000, the fleet grew quieter even though the number of airport operations increased.

In other cases, the challenge is to develop a future forecast noise contour that can withstand the challenge from sophisticated builders and their land use attorneys. This is no small task, and also one that is rife with risk to the municipalities, since a misstep by the planning agency could invite an inverse condemnation claim from a landowner. The uncertainty associated with calculations of noise contours is an issue for which current land use planning guidelines published by the FAA and state aviation agencies are weak, and planning and legal research is wanting. An unanswered legal question is whether such a "policy line" (in lieu of an explicitly defined noise contour) constitutes a commitment by the airport to manage operations to stay within that land use boundary. This could be construed as a de facto noise budget.

For example, if noise contours included an estimate of 95% confidence interval (i.e., a band within which the true contour value has a 95% chance of containment), the land use policy could be based on the confidence interval's outer boundary, inner boundary, or a mean contour location. Selecting such a boundary would be a policy, not a technical, decision. Attempting to make a "conservative" estimate would be difficult as the airport would see the outer band as the "conservative" while the developer would see the inner band as the "conservative" estimate.

7.2.8 *Modifications to Encourage Conduct of Social Surveys*

The prevalence of aircraft noise-induced annoyance depends not only on exposure levels, but also on community-specific non-acoustic factors (Fidell et al. 2011). The net effects of the latter variables cannot be measured with microphones or noise modeling software, but only via social survey. When predicting the prevalence of annoyance associated with proposed airport infrastructure projects, a generic disclosure of noise impacts (based on prospective modeling of exposure alone) can thus differ considerably from a community-specific disclosure, as discussed in Sect. 3.3. To increase the credibility of noise impact disclosures, FAR Part 150 and Master Planning exercises should be modified to fund routine conduct of social surveys to estimate Community Tolerance Level (CTL) values for communities affected by project-related noise exposure.

7.2.3 Mobilization to Encourage Conduct of Social Surveys

Acknowledgments

The authors are grateful to the many co-workers, clients, competitors, and professional colleagues, in academic and government organizations, as well as in consulting and law firms in the United States and abroad, who helped over many years to develop the understandings and influence the opinions expressed throughout this book. Comments on draft versions of this Monograph received from Geoffrey Gosling, Kevin Shepherd, and Steven Taber were particularly helpful.

S. Fidell and V. Mestre, *A Guide To U.S. Aircraft Noise Regulatory Policy*,
https://doi.org/10.1007/978-3-030-39908-5

Endnotes

1. The growth of the U.S. civil air transportation industry, initially encouraged by lucrative airmail contracts in the 1920s, was further subsidized by erection of high barriers to entry of new domestic competitors from the 1930s through the 1970s, as well as continuing cabotage protection from foreign competition; by creation of federally-sponsored airfields, navigation, and air traffic control systems; by post-World War II gifts of military training airfields to municipalities; by much federally-sponsored aeronautical research; by generous funding and favorable tax-advantaged financing for construction of airport infrastructure; by support for export markets for aircraft manufacturers; and so forth.
2. These include, *inter alia*, consideration of resources eligible for use in noise and emission reduction projects in a future aviation infrastructure and financing study; broadening the numbers of entities who can conduct assessments beyond airports; deadlines for completing FAA's ongoing evaluation of alternatives to the $L_{dn} = 65$ dB policy position and concluding the agency's review of the relationship between aircraft noise exposure and its effects on communities around airports; updating airport noise exposure maps "if in an area surrounding an airport, a change in the operation of the airport would establish a substantial new noncompatible use, or would significantly reduce noise over existing noncompatible uses"; and changing NextGen routing analyses and improvements to community involvement practices in metroplexes.
3. Initially, FAA's noise-related goals focused largely on promoting air transportation, facilitating airport capacity growth, and protecting public investment in aviation infrastructure. These goals later expanded to protecting the public from exposure to highly annoying aircraft noise, and mitigating the effects of such exposure. FAA (2012) acknowledges that despite its noise-related policies, "noise remains a predominant aviation environmental concern of the public, one of the principal environmental obstacles to expanding airport and airspace capacity, and the one that has used the most mitigation resources…".

S. Fidell and V. Mestre, *A Guide To U.S. Aircraft Noise Regulatory Policy*,
https://doi.org/10.1007/978-3-030-39908-5

4. It is more common than not for major runway construction and air traffic route-changing projects in the U.S. to be vigorously opposed by private interests, citizens' groups, and sometimes by county and municipal governments. Such organized opposition, which can add considerable cost and time to airport infrastructure development even when ultimately unsuccessful, has arisen, among other places, in Atlanta, Baltimore, Boston, Burbank, Chicago, Cincinnati, Denver, Los Angeles, Minneapolis, Newport Beach, New York, Oakland, Philadelphia, Phoenix, Pittsburgh, Reno, Sacramento, St. Louis, San Diego, San Francisco, Seattle, San Jose, and Washington, D.C. The problem is not unique to the U.S.; similar opposition is common in Australia, Canada, throughout Western Europe (especially in France, Germany, the Netherlands, and Scandinavia), and even in some Asian countries.
5. Two examples of white elephant airports are Lambert Field (STL) and Cincinnati/Northern Kentucky (CVG) airports. Both airports built infrastructure on the basis of unrealistic assumptions about future capacity needs. At STL, the assumption that TWA would continue to expand drove approval of a W-1W project to build a runway that proved to be superfluous after TWA collapsed (before the new runway was even completed!)
   At CVG, the airport expanded from one to four runways in the expectation that Delta would indefinitely continue its short- and long-haul hub operations. A mid-field terminal was also built in the expectation that Delta's subsidiary (Comair) would continue its regional hubbbing at CVG. Both expectations proved to be erroneous after Delta greatly reduced its hub operations at CVG. At both airports, independent assessments of the risks of over-estimating air traffic growth were lacking, and loose definitions of "significant" noise impacts contributed to mis-estimation of potential adverse consequences of airport expansion. At CVG, the excess runway capacity eventually encouraged expansion of an air cargo operation which brought new nighttime noise impacts to surrounding areas.
6. There are as many ways to measure noise as there are reasons for making the measurements in the first place. Detailed descriptions of aircraft noise measurement methods are beyond the scope of this monograph. Readers interested in transportation noise measurement may find a systematic overview of the topic in Mestre et al. (2011), among other places.
7. Day-Night Average Sound Level is a time-of-day and frequency-weighted measure of average noise exposure, normalized to a 24-hour period. It is abbreviated "DNL", and represented symbolically in mathematical expressions as L_{dn}. The measure was developed from a similar, pre-existing California noise metric ("CNEL") in a 1974 EPA document entitled "Levels of Noise Exposure Requisite to Protect Public Health and Welfare with an Adequate Margin of Safety".
8. FAR Part 36 certification limits on aircraft noise emissions are of no practical use in resolving actual airport/community noise controversies, since airports have no ability to deny airport access to a properly certificated but relatively noisy aircraft. In exceptional cases, an airport proprietor may shorten a runway

(as at Santa Monica Airport) after decades of negotiations with FAA, when it decides that an airport is *not* the highest and best use of valuable urban land, or even destroy a runway in a middle-of-the-night raid, as at (the former) Meigs Field in Chicago in 2003.

9. U.S. military laboratories began to sponsor and conduct research intended to advance the state of the art of aviation during World War I. During the World War II and Cold War years, much of the U.S. Air Force's in-house research on the effects of aircraft and rocket noise and vibration on aircrew was concentrated at the Wright Air Development Center (WADC) and its organizational successors. WADC, for example, sponsored development of the Community Noise Rating system, the first formal approach to prediction of community response to aircraft noise exposure.
   As summarized by Fidell (1979), "determination of a CNR value required estimating a "noise level rank" from a set of idealized spectral shapes for community noise. These shapes were derived from laboratory findings about the loudness of sounds in different frequency bands. The noise level rank was modified (normalized to standard conditions) by site-specific factors such as ambient noise levels, time of day and year, tonal content, dynamic range of noise intrusions, and novelty of exposure. CNR-based assessment of community reaction to environmental noise required a detailed case study, and involved more-or-less arbitrary judgments about the detailed nature of noise exposure".
   Under the direction of Dr. Henning von Gierke, the Air Force's Aerospace Medical Research Laboratory in Dayton, Ohio eventually extended its research efforts to include community response to aircraft noise, and played a dominant role in U.S. aviation noise research for more than two decades. Dr. von Gierke exerted a major influence in this process, arranging for development of airport noise modeling tools such as NOISEMAP software in the late 1960s (a decade before FAA showed much interest in aircraft noise modeling); recommending research activities for other government agencies; helping to organize the technical activities of the Office of Noise Abatement and Control of the Environmental Protection Agency under the Noise Control Act of 1972; and organizing a succession of voluntary federal interagency committees on urban and aircraft noise.
   Without any Congressional charter, the latter committees—FICUN, FICON, and FICAN—asserted self-proclaimed expertise for technical advice in aviation noise-related matters. Such expertise was intended to supplant that of the Committee on Hearing, Bioacoustics, and Biomechanics (CHABA) of the National Academy of Sciences, which had been the acknowledged authority in prior years. The inter-agency committees also closely coordinated policy positions of both military and civil agencies with interests in regulation of transportation noise. FAA adopted a similar strategy in later decades by establishing "Centers of Excellence" (assemblages of primarily academic institutions) to serve as docile channels for the conduct of FAA-funded research.
10. It may seem that outdoor residential noise exposure differs from personal noise exposure only by the sound transmission loss of the residence, but such is not

the case. Fidell et al. (2013a) have shown that when homes are occupied and residents are awake, noise levels in interior spaces of residences rarely differ just by a constant from exterior noise levels. On the contrary, interior noise levels are often controlled by the sounds of habitation, and vary greatly over the course of the day.

11. This position differs notably from one recommended by EPA in 1973. EPA (1973) recommends that "… any system used to characterize noise impact with respect to public health and welfare must be able to measure and/or calculate the noise exposure of individuals moving through different noise environments during their daily living routine. For example, occupational noise exposure during working hours, traffic noise during transportation to and from work, and the noise of the environment at home during evening and night all must be added to give the average noise level to which an individual is exposed during a day".
12. Stronger forms of epidemiological inference are based on individual-level information, as in case-control and longitudinal study designs. Without information about personal dosimetry, however, no such evidence is available to support regulatory analyses.
13. The locus of control of aircraft noise regulation remained unsettled through FAA's initial decades. The Aviation Safety and Noise Abatement Act of 1979 and the Airport Noise and Capacity Act of 1990 have since completely reversed the initial ("local control") view of aircraft noise regulation, substituting federal pre-emption for any effective local control over aircraft noise exposure by airports.
14. As an unintended consequence of modern technology (e.g., the advent of NextGen air traffic control and performance-based navigational routings, not to mention prospects for autonomous short range, low altitude urban flight operations, and for overland supersonic flight), adverse community reaction to aircraft noise may in the foreseeable future become more widespread and intense in areas remote from runway ends.
15. FAA does not use the term "significant" in its statistical sense of "unlikely to have occurred by chance alone". For all practical purposes, the term loosely implies "worthy (in FAA's judgment, for unclear reasons) of regulatory action". Originally (Rosenblith and Stevens 1953) the term "significant" implied levels of noise exposure likely to lead to unmanageable numbers of noise complaints at military airfields. The term currently implies a level of aircraft noise exposure that triggers mandatory consideration of mitigation alternatives and findings of impacts that require disclosure for NEPA purposes.
    Part 150 of the Federal Aviation Regulations ("FAR"), the 1985 implementing regulation for ASNA, completed the switch of FAA's preferred measures of noise exposure from CNR and NEF to DNL. (As explained in Sect. 2.1.8, the agency's former CNR = 100 and NEF = 30 dB criteria for significance were simply mathematical conversions into DNL units, without any consideration for the accompanying change in regulatory rationale from managing noise complaints to protecting some proportion of the population from exposure to highly annoying aircraft noise.)

The agency formally endorsed dosage-response analysis as the basis for its regulatory policy positions seven years after publishing FAR Part 150 (FICON 1992). FICON (Volume 2, Sect. 3.2.2.1, pp. 3–3 *et seq.*) states that "…the percent of the exposed population expected to be highly annoyed (%HA) [is] the most useful metric for characterizing or assessing noise impact on people"; and that "…the updated Schultz curve" remains the best available source of empirical dosage-effect information to predict community response to noise". FAA thus joined with other FICON member agencies in their adoption of a unique functional conversion of noise exposure into noise effects in exposed populations. In other words, dosage-response analysis transforms the effects of noise exposure into a monotonic function of DNL. Such a mathematical conversion also permits bi-directional translations of policy positions: from exposure units (decibels) to effect units, and from effect units (annoyance prevalence rates) to exposure units.

The particular dosage-response relationship that has served since 1992 as the ostensible rationale for FAA's definition of a threshold of "significance" associates a DNL value of 65 dB with a prevalence rate of high annoyance of 12.29% of the residential population. FAA has never published an explanation of why it considers 12.29% (as opposed to some other population proportion, perhaps 5 or 10%) as an appropriate threshold for regulatory action. Absent such an explanation, it remains unclear why a level of aircraft noise exposure that is expected to highly annoy an oddly arbitrary percentage of the population serves as a meaningful definition of "significant" noise exposure. As further described in Sect. 3.2.4, the dosage-response relationship which FAA cites as the basis for its analyses of aircraft noise impacts on residential populations is demonstrably obsolete and incorrect, and is inconsistent with the latest international technical consensus standard (ISO 2016).

16. FAA's understanding of "land use compatibility" reflects the agency's lack of a charter to foster municipal development. The term "compatibility" generally connotes some sort of reciprocity, so that (in this case) airports and communities would be compatible with one another. This is not the sense in which FAA understands land uses to be compatible with airports, however. From FAA's perspective, community land uses that are compatible with nearby airports are those which do not impair the continued operation and expansion of airports. FAA's working assumption is that the best and highest uses of land are always aviation-related, and that the proper goal of land use planning is invariably to protect existing public investment in airport infrastructure. Local land owners, developers, and municipalities (particularly those municipalities which are not airport proprietors) often disagree with this view, thus complicating compatible land use planning.

    FAA's definitions of land use compatibility do not take into consideration protection of public and private investment in non-aviation land uses. It is of little concern to FAA if aviation-related land uses are incompatible with community desires and intents, because FAA does not consider community development, protection of property tax bases of communities near airports, private property values, and the like to be among its responsibilities.

Although FAA does not acknowledge a direct duty under federal law to protect communities from aircraft noise exposure, it may be argued that it does in fact have such a duty. Section 2(b) of the Noise Control Act of 1972 states that "The Congress declares that it is the policy of the United States to promote an environment for all Americans free from noise that jeopardizes their health or welfare". Section 4(a) of the same act states that "The Congress authorizes and directs that Federal agencies shall, to the fullest extent consistent with their authority under Federal law administered by them, carry out the programs within their control in such a manner as to further the policy declared in Section 2(b)". Taken together, the two sections of the Noise Control Act would seem at the very least to broaden FAA's perspectives on "land use compatibility" to accommodate concerns other than threats to the continued operation and expansion of airports.

17. As late as publication of its Year 2000 Policy statement in the 14 July Federal Register, FAA articulated its view that it is the responsibility of others (not of the FAA, nor of airports), to protect airports from community development, rather than vice versa: "The FAA encourages local jurisdictions with responsibility for land use planning and zoning to take the strongest compatible land use actions within those areas around airports still subject to significant noise exposure…". In contrast, FAA's current vision statement, written since loss of its charter to promote civil aviation in 1996, now describes the agency as striving "to reach the next level of …environmental responsibility and global leadership" and accountability to the American public (*cf*. https://www.faa.gov/about/mission/).
18. In hindsight, ICAO's "balanced approach" appears to have been as much wishful thinking as a practical, internationally implementable policy position. Japanese law, for example, has no concept of eminent domain, so that a family farm was able for decades to prevent construction of an additional runway at the country's major international airport (Narita). Likewise, the concept of inverse condemnation (taking of the beneficial use and enjoyment of property without due process) is meaningless in Canada, where all land belongs to the Crown and is merely licensed to its current "owners".
19. The two most-recently created (and some speculate, last) major airports built on greenfield sites in the United States, Dallas-Ft. Worth (DFW) and Denver (DEN), were built on enormous sites. DFW was initially built on an18 square mile site, while DEN was built on a 50 square mile site. Older airports, many now landlocked and unable to expand, had often been built on sites as small as a few hundred acres.

In extreme cases, airport noise buffer zones constitute a form of "land banking", in which airports which cannot presently justify large exclusion areas for residential construction attempt to persuade zoning authorities that future regional economic growth and airport success will eventually require airport expansion. This is a form of a *lebensraum* social policy claim. Arguments of this sort raise fundamental questions about whether airports exist to serve communities, or communities exist to serve airports; about whether aviation-related land uses should or must invariably take precedence over all other land uses; and about which public and private institutions can or should control community development.

20. While some airports insulate residences of all sorts, most do not cover apartments (presumably because renters may frequently move) or condominiums (assumed to have fewer windows and better thermal insulation). The apartment assumption is not well supported in areas with rent controls, while the assumption of fewer windows in condominiums seems to be biased to colder climates, and may not be valid for temperate or hot climate airports.
21. A range of mandatory operational restrictions—on maximum permissible noise levels, on numbers of flights and hours of operation, on engine type, noise penalties, landing slots, and even complete noise budgets, have been attempted at various airports, but have survived at only a few, including DEN, JAC, JFK, LGB, SNA, and SMO. Federal pre-emption, via the Airport Noise and Capacity Act (ANCA) of 1990, brought an abrupt end to all such operational restrictions. The very effectiveness of ANCA in thwarting community efforts to impose operational constraints on airport operations has had unanticipated consequences however (Fidell and Mestre 2019). Local political opposition has now made it virtually impossible to construct a large, new civil airport on a greenfield site anywhere in the United States.
    Interestingly, it is not uncommon for an airport to be situated in a different political jurisdiction from that of its proprietor, and/or for the proprietor to be otherwise politically insulated from the population most impacted by its operations. Thus, for example, SFO, owned by the City and County of San Francisco, is located in San Mateo County, CA. ORD, owned by the City of Chicago, is situated outside of Cook County, IL. The largest population affected by noise of operations at CVG, owned by the Kenton County (Kentucky) Airport Board, is in Ohio. It can considerably ease pressures on airport proprietors when those who bear the noise burden of airport operations do not vote for public officials responsible for operating airports in other political jurisdictions.
22. Aviation interests have long lauded airports as "engines of economic development". It is therefore ironic for airport proprietors to appear surprised when magnets for economic development actually attract economic development to areas near airports. The issue of "encroachment" (of communities on airports, and of airport noise on communities near airports) is a particularly difficult one to resolve by federal fiat, since FAA has no jurisdiction over land use planning or uses of property that airports do not own.
23. The wording of Section 1113 of this legislation ("Study Concerning Relief of Homeowners in Proximity to Airports") is:

    > The Housing and Home Finance Administrator shall undertake a study to determine feasible methods of reducing the economic loss and hardship suffered by homeowners as the result of the depreciation in the value of their properties following the construction of airports in the vicinity of their homes, including a study of feasible methods

> of insulating such homes from the noise of aircraft. Findings and recommendations resulting from such study shall be reported to the President for transmission to the Congress at the earliest practicable date, but in no event later than one year after the date of the enactment of this Act.

24. Aircraft noise scales with engine power, which scales with aircraft weight. Setting noise limits by aircraft weight categories is an indirect, but effective expedient for setting noise limits for large jet transports.
25. Note that this definition conflicts with the Government Accounting Office's finding (2002) that officials from 35 airports report that "over half of their noise complaints during the last year came from people living in areas where aircraft noise falls below the level FAA considers incompatible with residential uses".
26. Today, airlines would almost certainly fault FAA for permitting airports to improve relations with communities by establishing airport-specific restrictions on aircraft types, on the grounds of inconsistency with the agency's charter to provide for an efficient national air transportation system, not to mention with ANCA. To a considerable degree, however, "efficiency" of a national air transportation system is in the eye of the beholder. Is de facto congestion pricing of airfares at over-scheduled airports more economically efficient on a national basis than increasing airline costs for serving certain airports only with specific aircraft types? Would not enforceable restrictions on the growth of airport operations yield fewer hours of inefficient flight delays nationwide, and/or facilitate construction of new airports? Is the degree of airline profitability a proper concern for decisions about the efficiency of a national air transportation system?
27. Non-acoustic factors that have been alleged to influence the annoyance of aircraft noise exposure include a wide range of personality traits (e.g., introversion, fearfulness), individual attitudes and characteristics (e.g., beliefs in misfeasance and malfeasance of airport proprietors), demographic factors such as age, gender, education, socioeconomic level, and situational variables (e.g., recent crashes, economic dependence on airport operation). Individual-level, non-acoustic factors are of little relevance for regulatory purposes, since aircraft fly over all neighborhood residents, without regard for their individual beliefs, sensitivities to noise, and circumstances. The *net* effects of non-acoustic factors in a given community, however, translate community-level dosage-effect relationships along the abscissa, as explained in Sect. 3.3.
28. Readers interested in a comprehensive and revealing account of the creation of Denver International Airport are referred to Dempsey et al. (1997).
29. Because the runway occupancy time for an aircraft takeoff or a landing at a busy civil airport is about two minutes, a single runway can accommodate no more than about 30 flight operations per hour under optimal conditions. Additional capacity constraints at multi-runway airports further reduce airport

capacity, especially in bad weather. These include converging or otherwise non-independent runways; runway-crossing terminal access routes, taxiway congestion, air traffic control limits to numbers of simultaneous approaches to independent runways, etc. Yet other factors, such as terminal capacity and even airport ground access, can also limit airport capacity, but do not affect aircraft noise exposure as directly or obviously as runway capacity (FAA 1983).

30. The year 2000 retirement of Stage II jet transports such as the B-727 from the fleet serving U.S. airports coincided closely with aircraft operators' economic self-interests. These included (1) lower labor costs of newer two-pilot cockpits (enabled by digital automation, which eliminated the position of flight engineer), and (2) lower direct operating costs due to the far greater fuel efficiency of later generation, high bypass ratio engines and larger airplanes.
31. For example, FAA has withheld publication of the findings of its 20-airport survey (Cointin et al. 2016) for years on the grounds that analyses of its findings remain incomplete.
32. FAA's Terminal Area Forecasts are uniformly optimistic, and not infrequently wrong.
33. Fidell and Mestre (2019) argue that this is a short-sighted perspective. Since the passage of ANCA abolished airport/community negotiation and compromise, the only practical recourse available to airport-vicinity communities which oppose airport growth is stonewall opposition to all such plans.
34. However much airports and regulatory bodies would prefer to characterize and deal with aircraft noise in purely acoustic terms, the problems that aircraft noise create remain stubbornly political and economic. The fact that annoyance is intangible, and cannot be directly measured with a sound level meter, does not make it any less real. Public opinion is also intangible, yet drives the entire democratic political process.

    Historically, the realization that aircraft noise effects cannot be successfully treated in exclusively physical terms dawned only slowly. As late as the 1950s, the World War II generation of acousticians construed adverse community reaction to aircraft noise as a largely behavioral (rather than an attitudinal) matter, related more or less directly to the frequency and extent of speech interference (which *can* be well-predicted from signal to noise ratios.) As increasingly larger and better-designed social surveys of airport neighborhood residents were conducted in later decades, it slowly became apparent that noise exposure alone controls only the lesser part of the variance from one community to the next in residential reactions to aircraft noise exposure. Formal acceptance of the systematic, joint influences of acoustic and non-acoustic factors on community reaction to aircraft noise was not achieved until the 2016 revision of ISO 1996-1.
35. The time constants of arousal and decay of annoyance (addressed in Sect. 3.4) are widely believed to be on the order of at least weeks or months. Time constants this long lend stability to annoyance prevalence estimates. In contrast, occurrences of speech interference are most usefully predicted from signal-to-noise ratios of short duration, discreet noise events. No systematic means exist to estimate or guide interpretation of community response to potential

speech interference-minutes, however. Simply because sound levels produced by hundreds of individual outdoor aircraft noise events sometimes exceed values that are potentially capable of reducing speech intelligibility to some degree implies little about actual speech interference instances. The actual extent of aircraft noise-induced speech interference in a community depends on many other situational factors: numbers of people at home at times of day coinciding with aircraft operations, who are actually trying to understand speech (whether in conversation, or by listening to telephone or other communication media), the ages, linguistic competence of both talkers and listeners, desired communicating distances, and the like.

36. FAA permits alternative measures of aircraft noise effects on communities to be considered on a case by case basis, but has never published any interpretive criteria for regulation of noise effects other than annoyance.
37. This preoccupation with complaints is a consequence of the lingering influence of CNR analysis on regulatory thinking, two decades after it was first developed, but prior to Schultz's (1978) *tour de force* meta-analysis of transportation noise annoyance.
38. Ironically, these 1970-era relationships come closer to contemporary European and international standards for assessing the prevalence of a consequential degree of aircraft noise-induced annoyance in communities than FICON's (1992) "updated" Schultz relationship.
39. Schultz's rationale for adopting an extreme (rather than average) definition of noise-induced annoyance was to avoid trivializing the concept of annoyance. Boeing (1973) had expressed doubts about the meaningfulness of "public annoyance", and others had questioned the notion of basing assessment of noise impacts in non-physical units. Today, such lingering doubts appear to be vestiges of the simplistic acoustic engineering approach to regulation of transportation noise prevailing in the 1950s and 1960s.
    Schultz's 1978 "synthesis" curve was the product of an "eyeball" polynomial fit, rather than a more formal statistical procedure. Generic software facilitating the conduct of multivariate correlational analyses of large data sets did not come into widespread use until the early 1980s. When advanced software for performing correlational analyses did become widely available, regression became the standard approach to analyzing social survey data on community reaction to aircraft noise exposure for the next several decades. It was not until the 2016 adoption of an international technical consensus standard (ISO 1996–1) that a causal, rather than a correlational, approach to defining and predicting relationships between noise exposure and annoyance, started to gain acceptance.
40. FICON's analysis omitted a number of aircraft noise data points from the analysis (for example, some from Burbank Airport, per Fidell et al. 1981) on the grounds that they appeared to be outliers. Following the subsequent conduct of social surveys at other airports, they no longer appear to be so.
41. Interestingly, FAA has never offered any such rationale for defining land use compatibility in units of decibels. Without such a rationale, the "compatibility" of a land use with the continued operation and expansion of airports is simply a

bureaucratic form of truth by assertion: an appeal to authority, unsupported by any systematic technical reasoning, and devoid of any intrinsic meaning. Land uses described by FAA as incompatible with airports are incompatible for no reasoned argument or explicit rationale other than FICUN's pronouncement in 1980 (a time when FAA's mission included promotion of civil aviation) that they are incompatible.

42. The FICON curve's underestimation of the annoyance of aircraft noise is due in large part to inclusion of field observations of the annoyance of road and rail noise in the FICON analysis. The degree to which the prevalence of high annoyance due to aircraft noise exposure exceeds that due to road and rail noise was not fully appreciated in the late 1980s when the database of field observations subsequently subjected to logistic regression to create the "updated Schultz curve" was assembled.
43. Few airport communities in the U.S. are exposed today to aircraft noise at levels greater than about $L_{dn} = 70$ dB. Aircraft noise exposure at levels lower than about $L_{dn} = 55$ dB in neighborhoods near runway ends of large airports is also essentially non-existent. As noted by Fidell and Silvati (2004), the supply of plausible regulatory action thresholds expressed at 5 dB intervals within this range of exposure levels is effectively limited to three: 65 dB (the current criterion), 60 dB, and 55 dB.
44. The loudness of a sound varies with both frequency and absolute level. The A-weighting network is insensitive to absolute sound pressure level, even though it was originally intended for relatively lower-level sounds.
45. The "annoyance decision criterion" is a scalar quantity that reflects the sum of all non-acoustic factors that influence decisions of survey respondents resident in the same communities to describe themselves as "highly annoyed" by noise exposure.
46. An ISO Technical Specification [2016a] offers procedural recommendations for the conduct of such surveys.
47. The choice of annoyance prevalence rate at which the CTL curve is anchored to the abscissa affects only a constant, is essentially arbitrary, and has no effect on the generality of the analytic approach.
48. Alternatively, Taraldsen et al. (2016) have suggested a maximum likelihood ratio method of fitting effective loudness functions to datasets of field observations. Estimates of CTL values made by the two (least square error and maximum likelihood) methods typically differ little (Gelderblom et al. 2017).
49. Military aviation can differ notably from the civil flight activity pattern. Flight operations at many military airfields tend to be a weekday, daytime activity, with relatively little weekend and night flying. DoD policy in the recent past acknowledged such differences by distinguishing Average Busy Day (i.e., weekday) and Average Annual Day noise modeling conditions. The distinction was made to avoid biasing (that is, underestimating) aircraft noise impacts by averaging days with little flight activity into annual average exposure estimates.

    More egregiously, at some naval airfields, Field Carrier Landing Practice exercises are periodically conducted as part of the training syllabus for visiting squadrons. Residents living near outlying (auxiliary) military airfields at such bases may be exposed several times a year to frequent low-altitude, touch-and-go flights by military jets at high engine power settings, often at night. At other times of year, such auxiliary airfields may experience little or no aircraft exposure for prolonged periods of time. The distinction between Average Busy Day and Average Annual Day noise modeling is far more acute in such cases.

50. As described by a U.S. Government Accounting Office report (2002), this view is reinforced by the observation that most aircraft noise complaints are lodged from areas with aircraft noise exposure levels which do not exceed the federally-defined threshold of significant noise exposure levels (L_{dn} > 65 dB) and/or incompatible land use. This is, of course, a circular argument that hinges on an arbitrary assertion that particular land uses are "incompatible" with noise exposure at levels of L_{dn} > 65 dB.
More recently, a July 12, 2013 decision of the Court of Appeals for the District of Columbia Circuit (No. 12-1335 in Helicopter Association International, Inc., Petitioner, v. Federal Aviation Administration, Respondent) confirmed that FAA has the authority to regulate flight paths on the basis of noise complaints outside the 65 dB DNL contour. The ruling indicates that FAA need not necessarily base its aircraft noise regulatory positions solely upon levels of aircraft noise exposure, but can also base them on documented aircraft noise complaints.
51. Although this finding is a familiar and long standing one, it has become politicized of late as a result of concerted efforts to disparage aircraft noise regulation. In a bid to denigrate aircraft noise regulation, the treatise by Dourado and Russell (2016) published by the externally-funded Mercatus Center at George Mason University (incorrectly) asserts that aircraft noise complaints play a meaningful role in federal aircraft noise policy.
52. This pattern of very large increases in complaint spatial density despite minimal changes in aircraft noise exposure suggests that complaints may be driven by proximity to flight paths (that is, more-or-less direct overflights), rather than by noise level per se, even though the two quantities are obviously highly correlated.
53. Behaviorally-confirmed awakening is the simplest, most direct and understandable, and hence most useful measure of sleep disturbance for purposes such as disclosure and understanding of environmental impacts of constructing airport infrastructure. Electrophysiologically- and motility-defined measures of sleep disturbance are more arcane than behavioral awakening, lack agreed criteria for interpretation, and have little or no experiential meaning for the public at large. Unfortunately, the recent retraction of ANSI/ASA S12.9-2008/Part 6 leaves no recognized, standard method for predicting behavioral awakening.
54. Anecdotal circumstances of susceptibility of highly skilled task performance to startling noise intrusions (e.g., diamond cutting or brain surgery) are easily imagined, but rarely occur in residential settings near runway ends.
55. Relative risk in epidemiologic research is a measure of the strength of association between a health effect and its putative cause. A relative risk, also called a risk ratio (RR), compares the risk of a health event (e.g., disease, injury, or death) in one group with the risk in another group. Relative risk is calculated by

dividing the incidence proportion in the first group by the incidence proportion in the second group. A relative risk thus represents a conditional probability—the ratio of disease incidence rates in variously exposed groups. For example, relative risk can quantify the probability of developing a disease for individuals exposed to aircraft noise relative to the probability of unexposed individuals' developing the disease.

In this context, a relative risk of 1.0 indicates no association between noise exposure and a health effect of interest. Relative risks only slightly different from 1.0 are commonplace in many epidemiologic studies of geographic associations between aircraft noise exposure and adverse health consequences. A relative risk greater than 1.0 implies a positive association, or increased risk, relative to a no- or low-exposure reference group. A relative risk smaller than 1.0 implies a negative association, suggesting that noise exposure might have a protective effect for a particular health consequence.

56. Recall the definition of noise in Chap. 1 as "sound that someone *else* considers too inconvenient to control".
57. It is a common fault (or feature, depending on perspective) of the democratic political process, and of Congress in particular, to simultaneously pursue mutually inconsistent goals. (In Churchill's words, "democracy is the worst form of Government except for all those other forms that have been tried from time to time…") U.S. politicians write conflicting goals into the charters of various executive branch agencies to satisfy their constituencies, then pronounce the underlying tensions solved. Doing so may de-politicize an issue for a while, but condemns the agencies so created to continued bureaucratic struggle with sister agencies.
58. Field studies designed to seek physical evidence of cortisol concentrations linkable to noise exposure to test the stress hypothesis of noise-induced disease are rare. One recent study (Michaud et al. 2016) concludes that "Multiple regression modeling left the great majority (77–89%) of the variance in perceived stress scale (PSS) scores, hair cortisol concentrations, resting blood pressure, and heart rate unaccounted for, and [wind turbine noise] exposure had no apparent influence on any of these endpoints".
59. FAA Order 5190.6B (2009) reserves airport revenue streams to airport capital and operating expenses, and precludes airport revenue from being "diverted" to compensate communities for aircraft noise-related expenses–even when they are direct consequences of airport operations.
60. Airports have proprietary views about the populations of "their" hinterlands. Airport business development offices commonly conduct surveys of passengers and of license plates of cars in other airports' parking lots to measure "leakage" of business from their catchment areas to nearby (competing) airports.
61. Light rail projects that have the potential for improving airport access are one example of areas in which airports may work with some communities, occasionally against the interests of other communities.
62. SFO's attorneys persuaded the jury in the appeal of the first round of small claims court awards to Superior Court that compliance with FAA standards

was all that the airport could do to control noise. Despite the Supreme Court decision in *Griggs* (*cf.* Sect. 2.1.3), the jury apparently believed that SFO was restricted by FAA policy and regulations from making changes to abate aircraft noise exposure.

63. Some state-level environmental disclosure legislation is more stringent than NEPA. For example, under California's Environmental Quality Act (CEQA), simple disclosure of an impact is insufficient. Any identified impacts must be addressed by appropriate mitigation, and mitigation monitoring must be conducted to confirm compliance. The disparity between a policy threshold of $L_{dn} = 65$ dB for the significance of noise exposure, and what many communities consider a tolerable level of noise exposure, is great enough that mitigation measures which reduce exposure to $L_{dn} < 65$ dB still leave many people highly annoyed.
64. Note that NEPA is purely procedural legislation. NEPA does not impose any liability on parties charged with implementing it, nor permit any recovery of damages for incorrect or misleading disclosures of environmental impacts. The legislation is intended only to produce a statement understandable by decision makers of adverse environmental consequences of proposed federal actions. An environmental disclosure document could comply with NEPA disclosure requirements even if it stated that a federally-sponsored action would wreak havoc on the environment. Further, local decision makers are free to approve a proposed project even if so advised.
65. To prevent ground reflections from influencing measured sound levels.
66. To prevent building reflections from influencing measured sound levels.
67. The confusion arises from the mathematics of computing averages for logarithmic quantities, such as decibels. The arithmetic average of 50 and 100 is 75. When the 50 and 100 values represent decibel values, however, they must first be converted into antilogs (by exponentiation, into their corresponding energy domain quantities) before averaging. When 50 dB and 100 dB are averaged in the energy domain, the resulting average is about 97 dB. In other words, higher noise level events contribute far more to DNL values than do lower noise level events. A noise monitoring system can provide estimates of DNL values adequate for common purposes by measuring only the louder events. The SAE document on airport noise monitoring systems (AIR 4721) and ISO 20906 provide best practice guidelines on noise event acquisition by monitoring systems.
68. The origins of computer-based aircraft noise modeling are poorly documented. Methods for creating hand-drawn aircraft noise exposure contours had been described by SAE Aerospace Recommended Practice 1114 ("Procedures for developing aircraft noise exposure contours around airports") in 1970, but had already been in existence for several years in the 1960s, as applied to characterizing aircraft noise exposure around U.S. Air Force bases. Taub et al. (1971) published a FORTRAN program for an airport noise model called Noise Exposure Model Mod-5 in 1971. Taub et al. credit the original version of the software to a model created for DOT by Serendipity, Inc. in 1971. In the same time frame,

however, pre-release versions of NOISEMAP had already been produced for the Air Force and applied at Barksdale and Castle Air Force bases, among other places.

69. In Chicago, for example, contractors wishing to participate in residential noise insulation projects at ORD were expected to support the municipal administration.
70. San Francisco International Airport (SFO) is one such example. The landlocked airport, situated on the shore of San Francisco Bay, is one of three large civil airports in the Bay Area. SFO lobbied intensively and invested heavily for years in efforts to fill enough of San Francisco Bay to construct one or more additional runways offshore. In the end, SFO's long-term growth plans were thwarted by overwhelming political opposition, largely on environmental grounds.
71. Airport costs also scale with number of enplanements, but airports have few other options for increasing revenues. Airports can also generally find uses for any excess funds for maintenance and facility upgrade purposes, and can always avoid federally-prohibited profits from aeronautical activity by refunding gate lease costs to airlines.
72. ICAO's "balanced approach" may be viewed as an elaboration of Beranek's (1978) familiar "Source-Path-Receiver" noise control model, particularized for airport applications. The noise source reduction, noise abatement procedures, and operating restrictions are directed at the source and path elements of the model, respectively, while the land use planning and management elements are directed at the receiver.

    ICAO'S balanced approach is grounded in a simplified acoustical engineering approach to characterizing management of airport noise impacts, in which noise exposure is the sole determinant of community reaction. The approach fails to deal effectively with the non-acoustic influences on annoyance prevalence rates, and thus ignores a potentially large part of the problem at some airports.
73. The term DNL is used here, whereas the actual metric used was CNEL, California's version of a cumulative time-weighted average noise level, incorporating both 5 dB evening (7:00 PM–10:00 PM) and a nighttime (10:00 PM–7:00 AM) penalty.
74. NextGen is FAA's term for the process in which air traffic control and navigation is being upgraded from ground-based 1950s technology to contemporary aircraft- and satellite based technology.
75. "We strive to reach the next level of safety, efficiency, environmental responsibility and global leadership. We are accountable to the American public and our stakeholders". (https://www.faa.gov/about/mission/).
76. The letter makes two further unsupported assertions. First, it claims that "… the existing 65 decibel day-night average sound level threshold and noise assessment procedures established under ASNA are supported by existing science" and "appropriately map the subjective experience of noise into a system that can be applied on an objective basis". In other words, the trade groups assert that FAA policy positions on the acceptability of aircraft noise exposure are supported by "existing science". In reality, FAA's regulatory policy positions

were developed long before publication of any modern dosage-response analysis; had nothing to do with dosage-response analysis of the sort endorsed by FAA and FICON; and were originally intended merely to keep aircraft noise complaints at military base housing within manageable limits.

It is beyond cavil that the "existing science" cited by the trade groups (that is, the "updated Schultz relationship" of the 1992 FICON report) is demonstrably biased, obsolete, and inconsistent with modern international technical consensus standards that predict the prevalence of annoyance due to aircraft noise exposure.

Second, the trade groups claim that "Airlines, other aircraft operators, airframe and aircraft engine manufacturers, and FAA have dramatically reduced the number of people exposed to aircraft noise...". This is a narrow and conditional claim that is plausible only with respect to an arbitrary definition of the level of aircraft noise exposure that FAA considers "significant". Although fewer people today than in the 1970s are exposed to levels of aircraft noise at levels in excess of $L_{dn} = 65$ dB, many more are exposed today to slightly lower levels of aircraft noise. As Schultz (1972) observed, is a slightly less obnoxious odor of rotting fish greatly preferable to a somewhat stronger odor of rotting fish?

The reasons that fewer people today are exposed to levels of aircraft noise in excess of $L_{dn} = 65$ dB than in the past are more closely linked to economic realities than to FAA's noise regulatory policies. The high bypass ratio engines that replaced the straight turbojet and turbofan engines of the early commercial jet fleet are far more fuel efficient than those of the first two generations of jet airliners, all of which were retired from the U.S. commercial fleet by 2000. It was the lower direct operating costs of aircraft with two-pilot cockpits (brought about by digital automation, not by federal noise regulatory policy) that fundamentally drove noisier aircraft out of the fleet.

77. U.S. courts reject policy positions adopted by regulatory agencies only when a case can be made that the positions are arbitrary and capricious.
78. An FAA-sponsored social survey of the annoyance of aircraft noise at twenty U.S. airports produced a draft final report in 2018. FAA has not at the time of this writing made the findings of the study public, nor announced any policy-related interpretations of them.
79. Freestone and Baker (2018), writing in an Australian context, acknowledge the negative externalities that airports impose on nearby communities even as they contribute to regional economic growth, but have little to suggest beyond some well-intentioned but impractical (in the American context, at least) measures to help resolve airport/community disputes. One of their suggested measures is creation of an ombudsman-like, impartial and non-political, body to mediate aircraft noise controversies. It is unclear how or why the usual parties to airport noise controversies in the U.S. would cede meaningful authority to such a body, and how it could come into existence
80. As a condition of membership, airport roundtables, noise compatibility committees, and the like typically require local governments to renounce flight path

changes that would shift aircraft noise exposure from one community to another. This stipulation, although superficially reasonable, reinforces the *status quo*, and precludes any possibility for sharing airport-environs noise burdens. Airport proprietors often control most important aspects of roundtable-like groups, such as membership, agendas, and meeting schedules. Airlines show little enthusiasm for participating in roundtable groups, particularly at fortress hubs, and have even resigned from them (for example, from the Minneapolis Metropolitan Aircraft Sound Abatement Council, MASAC) when they consider matters of voluntary or airline-specific modifications of flight schedules and fleet mixes.

81. The implicit rationale for expressing the significance of noise exposure in units of decibels is simplicity of interpretation. A decibel-valued definition of the significance of aircraft noise exposure, in combination with a set of aircraft noise exposure contours, lends itself to simple classifications of land use compatibility and eligibility for acoustic insulation and land acquisition. This rationale requires the further assumption, however, that all communities react identically to the same noise exposure, just as do sound level meters. This rationale is inconsistent with the great variability of annoyance prevalence rates among communities exposed to the same levels of aircraft noise.

    Expressing the significance of noise exposure indirectly (that is, in units of exposure, rather than in units of noise effect) also contributes to the common fallacy of misconstruing "the aircraft noise problem" simply as one of acoustic measurement. It is widely held that if a dosage-response relationship between DNL and the prevalence of annoyance with aircraft noise exposure doesn't fully explain community response to aircraft noise, perhaps a dosage-response relationship based on some other measure of aircraft noise might.

    The underlying fallacy associated with such superficial reasoning is the belief that acoustic measures of aircraft noise exposure exist that are sufficiently uncorrelated with DNL to yield a useful improvement in the predictability of annoyance prevalence rates. Such is not the case, however—all reasonable and useful measures of aircraft noise correlate so highly with DNL that they afford no meaningful improvement in variance accounted for in predictions of annoyance prevalence rates. The actual problem is that *non*-acoustic factors that can't be measured with a sound level meter or modeled by AEDT are responsible for much of the unexplained variance in predictions of annoyance prevalence rates.

Glossary

AC Abbreviation for Advisory Circular
AEDT Aviation Environmental Design Tool
ANCA Airport Noise and Capacity Act
ARP Aerospace Recommended Practice
ASNA Aviation Safety and Noise Act
CAB Civil Aeronautics Board, 1938–1985
CNEL Abbreviation for "Community Noise Equivalent Level", a California time of day- and frequency-weighted measure of cumulative noise exposure over a 24-hour period similar to DENL
CNR Community Noise Rating
CTL Abbreviation for "Community Tolerance Level", a value of DNL at which 50% of a community describes itself as highly annoyed by a transportation noise source
CVG International Air Transport Association three-letter airport identification code for Cincinnati/Northern Kentucky International Airport
DEN International Air Transport Association three-letter airport identification code for Denver International Airport
DENL Abbreviation for "Day–Evening–Night Average Sound Level", a time of day- and frequency-weighted measure of cumulative noise exposure over a 24-hour period
DHL A division of the German logistics company Deutsche Post DHL that operates express air cargo operations worldwide
DNL Abbreviation for "Day–Night Average Sound Level", a time of day- and frequency-weighted measure of cumulative noise exposure over a 24-hour period
DTW International Air Transport Association three-letter airport identification code for the Wayne County Airport Authority's Detroit Metropolitan Airport
ECAC Abbreviation for European Civil Aviation Conference
FAA Federal Aviation Administration
FAR Federal Aviation Regulation
FICAN Federal Interagency Committee on Aviation Noise
FICON Federal Interagency Committee on Noise

S. Fidell and V. Mestre, *A Guide To U.S. Aircraft Noise Regulatory Policy*,
https://doi.org/10.1007/978-3-030-39908-5

FICUN Federal Interagency Committee on Urban Noise

HHP International Air Transport Association three-letter airport identification code for Hong Kong airport

ICAO The International Civil Aviation Organization, a United Nations specialized agency formed in 1947 to codify the principles and techniques of international air navigation to foster the planning and development of international air transport

INM Integrated Noise Model: FAA's aircraft noise exposure prediction computer model, developed starting in the 1970s, and eventually incorporated into the agency's Aviation Environmental Design Tool software

JAC International Air Transport Association three-letter airport identification code for Jackson Hole Airport in Grand Teton National Park

JFK International Air Transport Association three-letter airport identification code for John F. Kennedy International Airport in New York City

ISO International Standards Organization

LAX International Air Transport Association three-letter airport identification code for Los Angeles International Airport

L_{ct} Symbol for CTL in mathematical expressions

L_{dn} Symbol for Day-Night Average Sound Level (DNL) in mathematical expressions

LGB International Air Transport Association three-letter airport identification code for Long Beach (California) Airport

MEM International Air Transport Association three-letter airport identification code for Memphis International Airport

NADP Abbreviation for Noise Abatement Departure Procedure, as described in AC 91-53A

NEF Abbreviation for Noise Exposure Forecast, a PNL-based measure of cumulative aircraft noise exposure

NEPA National Environmental Policy Act of 1970

OAK International Air Transport Association three-letter airport identification code for Oakland International Airport

ORD International Air Transport Association three-letter airport identification code for Chicago O'Hare International Airport

PBN Precision-based navigation

PFC Passenger facility charge—a form of head tax for transiting an airport that is consolidated in the price of airline tickets. PFC revenue for airports that is available for FAA-approved uses can be a major source of funding for airport infrastructure projects

PNL Abbreviation for perceived noise level, a loudness-like scale of individual noise event levels

RNP Required navigation procedure

RSIP Residential sound insulation project

SAE Society of Automotive Engineers

SAN International Air Transport Association three-letter airport identification code for San Diego International Airport

SEA International Air Transport Association three-letter airport identification code for Seattle-Tacoma International Airport

SFO International Air Transport Association three-letter airport identification code for San Francisco International Airport

SMO International Air Transport Association three-letter airport identification code for Santa Monica Municipal Airport

SNA International Air Transport Association three letter-airport identification code for John Wayne International Airport

Stages 1-5 Increasingly stringent aircraft noise emission limits established by FAR Part 36, corresponding to ICAO Chapters 1-5. The stage limits apply to aircraft weight categories, since aircraft weight scales with engine power, which in turn scales with noise. The stage limits establish maximum sound levels that may not be exceeded by aircraft offered for sale in the United States

STL International Air Transport Association three-letter airport identification code for St. Louis Lambert International Airport

VMC Visual Meteorological Conditions (also colloquially known as VFR, for Visual Flight Rules)

References

Airport Noise Report (ANR) (2018) "Noise grants totaling $116.8 million awarded to 16 airports in fiscal 2018" Volume 30, Number 38.

American Society of Testing and Materials (ASTM) (2018) "Standard Guide for Field Measurements of Airborne Sound Attenuation of Building Facades and Facade Elements" ASTM E966—18a.

Babisch, W., Houthuijs, D., Pershagen, G., Katsouyanni, K., Velonakis, M., Cadum, E., and Jarup, L. (for the HYENA consortium) (2008) "Hypertension and exposure to noise near airports—results of the HYENA study" 9th International Congress on Noise as a Public Health Problem (ICBEN).

Babisch, W., Houthuijs, D. Pershagen, G., Cadum, E., Katsouyanni, K., Velonakis, M., Dudley, H. Marohn, D., Swart, W., Breugelmans, O., Bluhm, G., Selander, J., Vignataglianti, F., Pisani, S., Haralabidis, A. (2009) "Annoyance due to aircraft noise has increased over the years—Results of the HYENA study" Environ. Int. 35 1169–1176.

Basner, M., and Samel, A. (2006) "Aircraft noise effects on sleep: application of the results of a large polysomnographic field study" J. Acoust. Soc. Am, 119(5), 2772–2784.

Basner, M., Babisch, W., Davis, A., Brink, M., Clark, C., Janssen, S., and Stansfeld, S. (2014) "Auditory and non-auditory effects of noise on health" Lancet, 383 (9925):1325–1332.

Bennett, R., and Pearsons, K. (1981) "Handbook of Aircraft Noise Metrics" National Aeronautics and Space Administration (NASA) Contractor Report 3406, NASA, Washington, DC.

Bennett, R., (1982) "Airport Noise Litigation: Case Law Review" Journal of Air Law and Commerce, Volume 47, https://scholar.smu.edu/jalc/vol47/iss3/4.

BENOX REPORT (prepared by H. W. Ades, W. R. Miles, H. Davis, W. D. Neff, D. H. Eldredge, I. Rudnick, H. E. von Gierke, A. A. Ward, W. C. Halstead, D. R. Warren, J. D. Hardy) (1953) "An Exploratory Study of the Biological Effects of Noise" Task Order 44 Office of Naval Research Project NR 144079, The University of Chicago.

Beranek, L. (1971) "Noise and vibration control", McGraw-Hill, New York.

Beranek, L. (2008) "Riding the Waves: A Life in Sound, Science, and Industry" MIT Press, Cambridge, MA.

Boeing (1973) Letter quoted in Appendix H of von Gierke, H. "Impact Characterization of Noise Including Implications of Identifying and Achieving Levels of Cumulative Noise Exposure,"U.S. Environmental Protection Agency, Report PB224408, Washington, D.C.

Booz-Allen Hamilton, Inc. (1994) "Effect of Airport Noise on Housing Values: A Summary Report. Prepared for the Federal Aviation Administration, Office of Environment and Energy.

Bolt Beranek and Newman (1960) "Planning guide for aircraft noise in residential areas, BBN Report 821, prepared for the Aeronautical Systems Division, U.S. Air Force Systems Command, Wright-Patterson AFB, Ohio.

S. Fidell and V. Mestre, *A Guide To U.S. Aircraft Noise Regulatory Policy*,
https://doi.org/10.1007/978-3-030-39908-5

Boon, B., and Wit, R. (2005) "The contribution of aviation to the economy: Assessment of arguments put forward" Solutions for Environment, Economy, and Technology, Delft, The Netherlands. Available from www.ce.nl.

Borsky, P. (1952) "Community aspects of aircraft noise" National Advisory Committee for Aeronautics.

Borsky, P. (1977) "A comparison of a laboratory and field study of annoyance and acceptability of aircraft noise exposures", NASA CR-27 72.

Brooker, P. (2009) "Do people react more strongly to aircraft noise today than in the past?" Appl. Acoust. 70 747–752.

California Code of Regulations, Title 21, State Noise Standards, 1971, revised 1990.

Kardous, C.A., and Shaw, P.B. (2016) "Evaluation of smartphone sound measurement applications (apps) using external microphones—A follow-up study" JASA Express Letters The Journal of the Acoustical Society of America 140, EL327.

City of Long Beach (1985) "Airport Noise Limits", Ordinance #C-7320, City of Long Beach Municipal Code Chapter 16.43.

Clarke, J.P. (2006) "Development, design, and flight test evaluation of a continuous descent approach procedure for nighttime operation at Louisville International Airport", Partnership for Air Transportation Noise and Emissions Reduction Report No. 2005-002.

Cointin, R., Sizov, N., and Hileman, J. (2016) "U.S. Civil Aircraft Noise Annoyance Survey Design," Internoise, Hamburg.

Cooley, J., and Tukey, J. (1965) "An algorithm for the machine calculation of complex Fourier series", Math. Comput. 19, 297–301.

Crocker, M. (1997) "Encyclopedia of Acoustics", Wyle-Interscience, ISBN 978-0471804659.

Dempsey, P. (2000) "Airport Planning and Development Handbook," McGraw Hill, ISBN0-07-134316-4.

Dempsey, P., Goetz, A., and Szyliowicz, J. (1997) "Denver International Airport: Lessons Learned," McGraw-Hill, New York, ISBN 0-07158184-7.

Dourado, E., and Russell, R. (2016) "Airport Noise NIMBYism: An Empirical Investigation" available at https://www.mercatus.org/system/files/dourado-airport-noise-mop-v1.pdf, last visited on 8/2/18.

Dykeman, P. (1997) "Speech to elected officials by the Deputy Director of FAA's Office of Environment and Energy (FAA/AEE) to elected officials, planners, and airport officials from jurisdictions near Raleigh-Durham International Airport on 16 September 1997," downloadable as of 14 June 2020 from the following URL: https://www.faa.gov/about/office_org/headquarters_offices/apl/noise_emissions/planning_toolkit/media/VI.C.pdf

Eagan, M. (2007) "Using Supplemental Metrics to Communicate Aircraft Noise Effects," U.S. Transportation Research Board, 2006 Annual Meeting CD-ROM.

Eagen, M., and Gardner, R., (2009) "Compilation of Noise Programs in Areas Outside DNL 65: A Synthesis of Airport Practice", Airport Cooperative Research Program, Transportation Research Board, Washington, D.C.

Economic Development Research Group, Inc., Gillen, D., ICF SH&E, Kramer Aerotek, Inc., and Mead & Hunt (2015) "The Role of U.S. Airportsin the National Economy" ACRP Report 132, Transportation Research Board, Washington, D.C.

Environmental Protection Agency (1973) "Impact characterization of noise including implications of identifying and achieving levels of cumulative noise exposure", NTID 73.4, Washington, DC.

Environmental Protection Agency (1974) "Information on Levels of Environmental Noise Requisite to Protect Public Health and Welfare with an Adequate Margin of Safety", Environmental Protection Agency (EPA) Report 550/9-74-094, EPA, Washington, D.C.

Environmental Protection Agency (2017) https://www.epa.gov/nepa/what-national-environmental-policy-act#NEPA, last visited on 4 February 2017. Environmental Protection Agency. (1974).

Eurocontrol (2002) Error Sensitivity Analysis of the Integrated Noise Model.

European Union (2002) "Directive relating to the assessment and management of environmental noise" Directive 2002/ 49/EC; see also European Commission, (2002) "Position paper on dose response relationships between transportation noise and annoyance", Luxembourg, Office for Official Publications of the European Communities.

Federal Aviation Administration (1983) "Airport Capacity and Delay" Advisory Circular 150/5060-5.

Federal Aviation Administration, "Noise Control and Compatibility Planning for Airports," (1983) Advisory Circular 150/5020-1.

Federal Aviation Administration, (2000) Docket No. 30109, 14 July 2000, "Aviation Noise Abatement Policy 2000", Federal Register.

Federal Aviation Administration (2012) "Aviation Environmental and Energy Policy Statement" www.faa.gov/about/office_org/headquarters_offices/apl/environ_policy_guidance/policy/media/FAA_EE_Policy_Statement.pdf

Federal Aviation Administration (2015) "Environmental Impacts: Policies and Procedures" Order 1050.1F, FAA/AEE, Washington, DC.

Federal Aviation Administration (2008), "FAA Integrated Noise Model Validation: Analysis of Air Carrier Flyovers at Seattle-Tacoma Airport (*sic*)".

Federal Aviation Administration (2009) "Airport Compliance Manual Order 5190.6B" Chapter 15, available at https://www.faa.gov/airports/resources/publications/orders/compliance_5190_6/media/5190_6b_chap15.pdf.

Federal Aviation Administration (2017) "Aviation Environmental Design Tool Version 2b Uncertainty Quantification Report".

Federal Aviation Administration (2019) Airport Noise Compatibility Planning Toolkit," www.faa.gov/about/office_org/headquarters_offices/apl/ /noise_emissions/planning_toolkit/.

Federal Aviation Administration (2019) Aircraft Noise Issues Webpage, https://www.faa.gov/about/office_org/headquarters_offices/apl/noise_emissions/airport_aircraft_noise_issues/.

Federal Interagency Committee on Noise (FICON) (1992) "Federal Agency Review of Selected Airport Noise Analysis Issues.

Federal Interagency Committee on Urban Noise (FICUN) (1980) "Guidelines for Considering Noise in Land Use Planning and Control" US Government Printing Office Report No. 1981–337–066/8071, Washington, DC.

Fidell, S. (1978) "Nationwide Urban Noise Survey," J. Acoust. Soc. Am., Vol. 64(1), 198–206.

Fidell, S., (1979) "Community Response to Noise," Chapter 36 of Handbook of Noise Control, Second Edition, C. Harris, Ed., McGraw-Hill, Inc.

Fidell, S. (2003) "The Schultz curve 25 years later: A research perspective" Journal of the Acoustical Society of America 114, 3007–3015.

Fidell, S. (2014) "Psychoacoustics and community noise impact" Acoustics Today 10, 40–44.

Fidell, S. (2015) "A Review of U.S. Aircraft Noise Regulatory Policy," Acoustics Today, 11 (4), 26–34.

Fidell, S. (ed.) (2018) "Rationale for Withdrawing ANSI/ASA S12.9-2008/Part 6" ASA TR S12.9-2018/Part 6.

Fidell, S., Barber, D., and Schultz, T. (1991) "Updating a dosage-effect relationship for the prevalence of annoyance due to general transportation noise," J. Acoust. Soc. Am. 89(1), 221–233.

Fidell, S., Horonjeff, R., Mills, J., Baldwin, E., Teffeteller, S., and Pearsons, K., (1985) "Aircraft Noise Annoyance at Three Joint Air Carrier and General Aviation Airports" J. Acoust. Soc. Am., 77(3), pp. 1054-1068. (See also Fidell, S. and Pearsons, K., "Comments on 'The Effect of Changes in Aircraft Noise Exposure," 1985 J. Sound Vib., Vol. 103, No. 1, pp. 139–140.)

S. Fidell, R. Horonjeff, S. Teffeteller, and K. Pearsons (1981) "Community Sensitivity to in Aircraft Noise Exposure" NASA Langley Research Center, Contract NASl-14611.

Fidell, S., and Mestre, V. (2019) "ANCA: Short term gain leads to long term pain for airport capacity enhancement", Noise-Con 2019, San Diego, CA.

Fidell, S., Pearsons, K.S., Tabachnick, B., Howe, R., Silvati, L., and Barber, D. (1995) "Field Study of Noise-Induced Sleep Disturbance," J. Acoust. Soc. Am., Vol. 98, No. 2, Pt. 1, pp. 1025–1033.

Fidell, S., Schultz, T. J., and Green, D. M. (1988). "A theoretical interpretation of the prevalence rate of noise-induced annoyance in residential populations." Journal of the Acoustical Society of America 84, 2109–2113.

Fidell, S., Tabachnick, B., and Silvati, L. (1996) "Effects of Military Aircraft Noise on Residential Property Values", BBN Report 8102.

Fidell, S., Mestre, V., Schomer, P., Berry, B., Gjestland, T., Vallet, M., and Reid, T. (2011). "A first-principles model for estimating the prevalence of annoyance with aircraft noise exposure" Journal of the Acoustical Society of America 130, 791–806.

Fidell, S., Mestre, V., and Sneddon, M., (2012) "A potential role for noise complaints as a predictor of the prevalence of annoyance with aircraft noise," Noise Control Eng. J. 60(1), 62–68.

Fidell, S., Mestre, V., Schomer, P., Horonjeff, R., and Reid, T. J. (2014) "A systematic rationale for defining the significance of aircraft noise impacts." Journal of the Acoustical Society of America 136, 1129–1138.

Fidell, S., and Mestre, V. (2019) "ANCA: Short term gain leads to long term pain for airport capacity enhancement", NOISE-CON 2019, San Diego, CA

Fidell, S., and Pearsons, K. (2003) "Sensitivity to prospective transportation noise exposure", Noise Control Eng. J., 51(2), 106–111.

Fidell, S., and Silvati, L. (2004) "Parsimonious alternatives to regression analysis for characterizing prevalence rates of aircraft noise annoyance." Noise Control Engineering Journal 52, 56–68.

Fidell, S., Sneddon, M., and Harrison, R., (2013a) "Relative contributions of highway and neighborhood sources to outdoor and indoor residential noise," Noise Control Eng. J. 61(2).

Fidell, S., Sneddon, M., and Silvati, L. (2001) "Review of Calendar Year 2000 Noise Exposure Contours for O'Hare International Airport," Fidell Associates Report 0103.

Fidell, S., Tabachnick, B., Mestre, V., and Fidell, L. (2013b) "Aircraft noise-induced awakenings are more reasonably predicted from relative than from absolute sound exposure levels," J. Acoust. Soc. Am. 134(5), (2013), 3645–3653.

Fidell, S., Teffeteller, S., Horonjeff, R., and Green, D., (1979) "Predicting Annoyance from Detectability of Low Level Sounds,"J. Acoust. Soc. Am., Vol. 66, No. 5, 1427–1434.

Fields, J. M. (1993). "Effect of Personal and Situational Variables on Noise Annoyance in Residential Areas," J. Acoust. Soc. Am. 93, 2753–63.

Fields, J. and Hall, F. (1987) "Community Effects of Noise." In Transportation Noise Reference Book, edited by P.M. Nelson. Cambridge, Great Britain: Butterworth.

Fields, J., (1984) "The effect of numbers of noise events on people's reactions to noise: An analysis of existing survey data", J. Acoust. Soc. Am. 75(2).

Freeman, A., and Farris, J. (1992) "Grassroots Impact Litigation: Mass Filing of Small Claims" University of San Francisco Law Review 261–281.

Freestone, R., and Baker, D. (2018) "Negotiating the Problem of Airport Noise: Comparative Lessons from the Australian Experience" The Political Quarterly, 1-9, John Wiley & Sons, Ltd. Oxford OX4 2DQ, UK.

Garner, W., (1947) "The effect of frequency spectrum on temporal integration of energy in the ear" J. Acoust. Soc. Am. 19, 808–815.

Galloway, W., and Pietrasanta, A. (1963) "Land Use Planning with Respect to Aircraft Noise." AFM 86-5, TM 5-365, NAVDOCKS P-98, Department of Defense, Washington, DC. Also published by the Federal Aviation Administration (FAA) as TR-821, October 1964.

Glass, D., and Singer, J. (1972) "Urban Stress: Experiments on Noise and Social Stressors" Academic Press, New York.

Gelderblom, F., Gjestland, T., Granøien, I., Taraldsen, G. (2014) "The impact of civil versus military aircraft noise on noise annoyance." Proc. INTER-NGarnerOISE 14, Melbourne, Australia, 1–10.

Gelderblom, F., Gjestland, T, Fidell, S., and Berry, B. (2017) Acta Acustica United with Acustica, Volume 103, Number 1, 17–27(11).

Gjestland, T. (2018) "A Systematic Review of the Basis for WHO's New Recommendation for Limiting Aircraft Noise Annoyance", Int. J. Environ. Res. Public Health, 15(12), 2717.

Green, D. M., and Fidell, S. (1991) "Variability in the criterion for reporting annoyance in community noise surveys," J. Acoust. Soc. Am. 89, 234–243.
Guski, R. (2004) "How to forecast community annoyance in planning noisy facilities." Noise Health 6, 59–64.
Government Accounting Office (2002) "Aviation and the environment: results from a survey of the nation's 50 busiest commercial service airports", GAO/RCED-00-222 35–36.
Green, D.M., and Fidell, S. (1991) "Variability in the criterion for reporting annoyance in community noise surveys," Journal of the Acoustical Society, Vol 69, 1991, pp. 234–243.
Hall, F., Birnie, S., Taylor, S., and Palmer, J. (1981) "Direct comparison of community response to road traffic noise and to aircraft noise", J. Acopust. Soc Am. 70(6), 1690–1698.
Harper, D. (1988) "Regulation of Aircraft Noise at Major Airports: Past, Present, and Future" Transportation Law Journal, 17, Sturm College of Law, Denver, CO.
Hill, A.B. (1965) "The environment and disease: association or causation?" Proc. R. Soc. Med. 58:295–300.
Hooper, P., Maughan, J., Flindell, I., and Hume, K. (2009) "Indices to enhance understanding & management of community responses to aircraft noise exposure" OMEGA Community Noise Study, 550/9-74-004.
Horonjeff, R., and Robert, W. (1997) "Attitudinal responses to changes in noise exposure in residential communities." NASA Contractor Report CR-97-205813.
Housing and Urban Development, Department of (HUD), "Noise Assessment Guidelines," August 1971 (Also published by Schultz, Theodore, BBN Report 2176, August 1971).
Indianapolis Business Journal, "Airport Authority sells 103 homes after sound modifications," July 17, 2006.
International Civil Aviation Organization (1982) "The balanced approach to aircraft noise measurement", ICAO Document 9829. AN/451.
International Civil Aviation Organization (ICAO), (2008) "Recommended Method for Computing Noise Contours Around Airports," Doc 29, 2008.
International Electro-Technical Commission (IEC), "Electroacoustics—Sound level meters—Part 1: Specifications," Standard 61672, 2013.
International Organization for Standardization (ISO) (2016a) "Assessment of noise annoyance by means of social and socio-acoustic surveys", ISO Technical Specification 15666, Geneva, Switzerland.
International Organization for Standardization (ISO) (2016b). Acoustics – Description, Measurement and Assessment of Environmental Noise - Part 1: Basic Quantities and Assessment Procedures, ISO 1996-1, Geneva, Switzerland.
International Organization for Standardization (ISO) (2009) "Acoustics—Unattended monitoring of aircraft sound in the vicinity of airports".
Janssen, S., Vos, H., Van Kempen, E, Breugelmans, O,, and Miedema, H. (2011) "Trends in aircraft noise annoyance: the role of study and sample characteristics". J. Acoust. Soc. Am. 129, 1953–1962.
Janssen, S. and Guski, R (2015) "Aircraft noise and health: Review of evidence." Directorate General Joint Research Center and Directorate General for Environment, European Union, Chapter 7.
Job, R. F. S. (1988) "Community response to noise: A review of factors influencing the relationship between noise exposure and reaction," J. Acoust. Soc. Am. 83, 991–1001.
John Wayne Airport, Orange County (2017) "Noise Abatement Program Quarterly Report for July 1, 2017 through September 30, 2017, https://www.ocair.com/reportspublications/AccessNoise/noiseabatementquarterly/2017/2017-q3.pdf.
Kastka, J., Borsch-Galetke, E., Guski, R., Krauth, J., Paulsen, R., Schümer, R., Oliva, C., (1995) "Longitudinal study on aircraft noise effects at Düsseldorf airport 1981–1993" Proc. ICA 95, Trondheim, 447–451.
Kryter, K. (1984) "Physiological, psychological, and social effects of noise," NASA Reference Publication 1115.

Kuhn, T. (1962) "The Structure of Scientific Revolutions", The University of Chicago Press, ISBN-13: 978-022645812.

Liva, C. (1998) "Belastungen der Bevölkerung durch Flug- und Strassenlärm: Eine Lärmstudie am Beispiel der Flughäfen Genf und Zürich." (People's exposure to aircraft and road traffic noise: An example of a noise study at Geneva and Zurich Airports). Duncker & Humboldt, Berlin.

McKennell, A. C. (1963) "Committee on the Problem of Noise: Final Report." Her Majesty's Stationary Office, London.

Mestre, V., Schomer, P., Fidell, S., and Berry, B. (2011) "Technical Support for Day/Night Average Sound Level (DNL) Replacement Metric Research" Purchase Order DTRT57-10-P-80191, Requisition No. DTRT-RVT-41-1113, Report Number DOT/FAA/AEE/2011-02, USDOT/RITA/Volpe Center, US Department of Transportation, Washington, DC.

Michaud, D., Fidell, S., Pearsons, K., Campbell, K., and Keith, S. (2007) "Review of field studies of aircraft noise and sleep disturbance," J. Acoust. Soc. Am. 121 (1), 32–42.

Michaud, D., Feder, K., Keith, S., Voicescu, S., Marro, L., Than, J., Guay, M., Denning, A., Bower, T., Villeneuve, P., Russell, E., Koren, G., and van den Berg, F. (2016) "Self-reported and measured stress related responses associated with exposure to wind turbine noise." J Acoust. Soc. Am. 2016 Mar; 139(3):1467–1479.

Miedema, H., and Vos, H. (1998) "Exposure-response relationships for transportation noise" Journal of the Acoustical Society of America, 104, 3432–3445.

Miedema, H., and Vos, H. (1999) "Demographic and attitudinal factors that modify annoyance from transportation noise" J. Acoust. Soc. Am. 105, 3336.

Miedema, H., and Oudshoorn, C. (2001) "Annoyance from transportation noise: relationships with exposure metrics DNL and DENL and their confidence intervals" Environmental Health Perspectives 109(4):409–16

Miller, N., Anderson, G., Horonjeff, R., Kimura, S., Miller, J., Senzig, D., Thompson, R. (2000) "Examining INM Accuracy Using Empirical Sound Monitoring and Radar Data" NASA Contractor Report 2000-210113.

Miller, N., Sizov, N., Lohr, S., and Cantor, D. (2014) "New research on community reaction to aircraft noise in the United States", 11th International Conference on Noise as a Public Health, Nara, Japan.

Morrison, S., Winston, C., and Watson, W. (1999) "Fundamental Flaws of Social Regulation: The Case of Airplane Noise" *The Journal of Law & Economics*. Vol. 42, No. 2 (October 1999), pp. 723–744, University of Chicago Press.

Nelson, J. (1978) "Aircraft Noise and the Market for Residential Housing: Empirical Results for Seven Selected Airports", DOT Report DOT-RSPA-DPB 50-78.

Nelson, J. (2004) "Meta-Analysis of Airport Noise and Hedonic Property Values: Problems and Prospects" Journal of Transport Economics and Policy, Volume 38, Part 1, January 2004, pp. 1–28.

Plotkin, K. J., Bradley, K. A., and Hobbs, C. M. (1998) "Examination of the lateral attenuation of aircraft noise," Wyle Laboratories Report WR 99–10.

Rosenblith, W. A., Stevens, K. N., and the Staff of Bolt, Beranek, and Newman (1953) Handbook of Acoustic Noise Control. Volume II. Noise and Man. WADC TR-52-204, Wright Air Development Center, Wright-Patterson Air Force Base, Ohio.

Sabel, J. (2004) "Airline-Airport Facilities Agreements: An Overview" Journal of Air Law and Commerce, 69(4).

Schomer, P., Mestre, V., Fidell, S., Berry, B., Gjestland, T., Vallet, M., and Reid, T. (2012) Role of community tolerance level (CTL) in predicting the prevalence of the annoyance of road and rail noise. Journal of the Acoustical Society of America 131, 2772–2786.

Schultz, T. J. (1972). Community Noise Rating. Applied Science Publishers, London.

Schultz, T. J. (1978). Synthesis of social surveys on noise annoyance. Journal of the Acoustical Society of America 64, 377–405.

Shapiro, S. (1992) Lessons from a Public Policy Failure: EPA andNoise Abatement 19 Ecology L. Q. 1 Available at: http://scholarship.law.berkeley.edu/elq/vol19/iss1/1.

Smith, M.J.T. (1989) "Aircraft Noise", Cambridge Aerospace Series, Cambridge University Press, ISBN 0521616999.

Society of Automotive Engineers (SAE), (2006) "Monitoring Aircraft Noise and Operations in the Vicinity of Airports: System Description, Acquisition, and Operation," ARP 4721.

Stevens, K. N., Rosenblith, W., and Bolt, R. (1955) A community's reaction to noise: Can it be forecast? Noise Control 1, 63–71.

Stevens, K., Pietrasanta, A., and the Staff of Bolt Beranek and Newman, Inc." (1957) "Procedures for estimating noise exposure and resulting community reactions for air base operation", WADC TN-57-10, Wright-Patterson Air Force Base, Ohio (Wright Air Development Center).

Suter, A (1991) "Wendell Ford's Edsel—or How to Delight the Lobbyists and Enrage the Citizens" Sound and Vibration.

Taub, J., Foreman, T., and Brownfield, B. (1971) "The Noise Exposure Model MOD-5", Report No. DOT-TSC-OST-72-5, Transportation Systems Center.

Thompson, S., Fidell, S., and Tabachnick, B. (1989) Feasibility of Epidemiologic Research On Non-auditory health effects of Residential Aircraft Noise Exposure, U.S. Air Force Technical Report HSD-TR-89-007. (See also Thompson, S. and Fidell, S., (1988), Feasibility of Studying Human Health Effects of Aircraft Noise in Residential Populations In: Berglund, B., and Lindvall, T., Eds., Noise as a Public Health Problem, Vol. 4, Swedish Council for Building Research, Stockholm, Sweden.)

Tracor (1971) "Community reaction to airport noise—Volume 1", NASA Contractor Report CR-1761.

U.S. Department of Transportation (2016) "The Economic Impact of Civil Aviation on the U.S. Economy" available at https://www.faa.gov/air_traffic/publications/media/2016-economic-impact-report_FINAL.pdf.

Van Kempen, E., Van Kamp, I. (2005) "Annoyance from air traffic noise: Possible trends in exposure-response relationships." RIVM National Institute for Public Health and the Environment, 01/2005 MGO EvK.

Volpe Transportation Research Center (2010) "A re-analysis of Day-Night Sound Level (DNL) as a function of population density in the United States," Office of the Assistant Secretary for Transportation Policy, Office of the Secretary, United States Department of Transportation, February 2010.

von Gierke, H. (1973) "Impact Characterization of Noise Including Implications of Identifying and Achieving Levels of Cumulative Noise Exposure," U.S. Environmental Protection Agency, Report PB224408, Washington, D.C.

von Gierke, H. (1979) in Harris, C. "Handbook of Noise Control, Second Edition" Chapter 45, "Environmental Impact Statements", McGraw-Hill Book Company, New York.

Waitz, I., Townsend, J., Crutcher-Gershenfeld, J., Greitzer, E., and Kerrebrock, J. (2004) "Report to the United States Congress: Aviation and the Environment - A National Vision Statement, Framework for Goals and Recommended Actions", Partnership for AiR Transportation Noise and Emissions Reduction.

Waitz, I., Bernhard, R., and Hanson, C. (2007) "Challenges and Promises in Mitigating Transportation Noise" The Bridge, National Academy of Engineering, Volume 37, Number 3.

Weiner, I., Tennen, H., and Suls, J. (2012) Handbook of Psychology, Personality and Social Psychology, Wiley, New York.

Wilson, D. K., Valente, D., Nykaza, E., and Pettit, C. (2013). Information-criterion based selection of models for community noise annoyance. Journal of the Acoustical Society of America 133, EL195-EL201.

Woodward, J., Briscoe, L., and Dunholter, P. (2009) "Aircraft Noise: A Toolkit for Managing Community Expectations" Chapter 6, Aircraft Cooperative Research Program Report 15, National Academy of Science, Washington, D.C.

Wyle Laboratories, (1971) "Supporting Information for the Adopted Noise Regulations for California Airports: Final Report for California Department of Aeronautics", Wyle Report WCR 70-3

(Revised); See also "Community Noise," a Report prepared for Office of Noise Abatement and Control, U.S. Environmental Protection Agency, under Contract 68-04-0046.

Zur Nieden, A., Römer, K., Spilski, J., Möhler, U., Harpel, S., Schreckenberg, D., Eikmann, T. (2016) "NORAH - field study: The effects of chronic exposure to traffic noise (aircraft, rail and road) on hypertension," InterNoise 16, Hamburg, Germany, 1879-2861, 2201–2204.

Zwicker, E., and Fastl, H. (1999) Psychoacoustics: Facts and Methods, 2nd Edition. Springer-Verlag, Berlin.

The manufacturer's authorised representative in the EU is Springer Nature Customer Service Centre GmbH, Europaplatz 3, 69115 Heidelberg, Germany. If you have any concerns regarding our products, please contact ProductSafety@springernature.com

Printed and bound by CPI Group (UK) Ltd, Croydon, CR0 4YY

30/09/2026

02225338-0003